Literature and sustainability

Manchester University Press

In memory of John Parham
(1964–2023)

Literature and sustainability

Concept, text and culture

Edited by

Adeline Johns-Putra, John Parham

and **Louise Squire**

Manchester University Press

Published by Manchester University Press
Oxford Road, Manchester M13 9PL
www.manchesteruniversitypress.co.uk

British Library Cataloguing-in-Publication Data
A catalogue record for this book is available from the British Library

ISBN 978 0 7190 9967 0 hardback
ISBN 978 1 5261 8235 7 paperback
ISBN 978 1 5261 0763 3 open access

First published 2017
Paperback published 2024

Typeset
by Toppan Best-set Premedia Limited

Contents

List of figures vii

List of contributors viii

Foreword by Gillen D'Arcy Wood xii

Acknowledgements xvi

Editors' introduction – Adeline Johns-Putra, John Parham and Louise Squire 1

PART I: Discourses of sustainability **11**

1 The millers' tales: sustainability, the arts and the watermill – Jayne Elisabeth Archer, Howard Thomas and Richard Marggraf Turley 13

2 Sustenance from the past: precedents to sustainability in nineteenth-century literature and culture – John Parham 33

3 Deep sustainability: ecopoetics, enjoyment and ecstatic hospitality – Kate Rigby 52

4 Recycling materials, recycling lives: cardboard publishers in Latin America – Lucy Bell 76

5 Sustainability after extinction: on last animals and future bison – Joshua Schuster 97

6 The twilight of the Anthropocene: sustaining literature – Claire Colebrook 115

PART II: Reading sustainability **137**

7 Collapse, resilience, stability and sustainability in Margaret Atwood's MaddAddam trilogy – Dana Phillips 139

8 'The shadow of the future made all the difference': sustainability in Kim Stanley Robinson's Science in the Capital trilogy – Chris Pak 159

9 The unsustainable aesthetics of sustainability: the sense of an ending in Jeanette Winterson's *The Stone Gods* – Adeline Johns-Putra 177

10 A modest proposal for a less natural lifestyle: the paradoxes of sustainability and Michel Houellebecq's *The Possibility of an Island* – Hannes Bergthaller 195

11 Jorie Graham's *Sea Change*: the poetics of sustainability and the politics of what we're sustaining – Matthew Griffiths 211

12 Circles unrounded: sustainability, subject and necessity in Yann Martel's *Life of Pi* – Louise Squire 228

Index 246

Figures

1.1 Illustrated frontispiece, William Camden, *Britannia, or A Chorographicall Description of the Most Flourishing Kingdomes, England, Scotland, and Ireland, and the Ilands Adioyning*, trans. Philemon Holland (London: Georgii Bishop & Ioannis Norton, 1610). Welcome Library, London CC BY 4.0 16
1.2 John Constable, *The Hay Wain* (1821), oil on canvas, 130.2 cm × 185.4 cm, National Gallery (NG1207) © The National Gallery, London 20
1.3 Felin Ganol, Llanrhystud, Ceredigion © 2015 Anne and Andrew Parry, Felin Ganol, Llanrhystud, Wales (www.felinganol.co.uk), used with permission 27
4.1 Publishers of Dulcinéia Catadora: Maria Dias, Andréia Ribeiro, Elizângela Juventino, Eminéia Silva Santos, used with permission 85
4.2 Page design of *Catador*, including photograph of torn cardboard as margin, from *Catador* [*Waste Picker*] (São Paulo: Dulcinéia Catadora, 2012), 16, used with permission 86
4.3 The waste pickers' cart transformed by Dulcinéia Catadora into a stall for their beautiful cardboard books at the third Feira Plana (São Paulo, 2014). Photograph by Marcos Rosa (son of founder of Dulcinéia, Lucía Rosa) 92
5.1 Adrian Stimson, *Fuse 2* (2010) © Adrian Stimson 2010, reproduced by permission of the artist 110
5.2 Kent Monkman, *The Chase* (2014) © Kent Monkman 2014, reproduced by permission of the artist 111

Contributors

Jayne Elisabeth Archer teaches Renaissance and nineteenth-century literature at the University of Bedfordshire. Her research interests include Shakespeare, the court and culture of Queen Elizabeth I, and the relationship between science, magic and literature. She has published the award-winning five-volume *John Nichols's The Progresses and Public Processions of Queen Elizabeth I* (2014), *Food and the Literary Imagination* (2014), *The Intellectual and Cultural World of the Early Modern Inns of Court* (2011) and *The Progresses, Pageants, and Entertainments of Queen Elizabeth I* (2007).

Lucy Bell is Lecturer in Spanish and Translation Studies at the University of Surrey. She completed her PhD at the University of Cambridge in 2013. She has published widely in literary studies, critical theory, Latin American studies and environmental criticism. Recent articles include 'Understanding Concepts of Locality, Resistance, and Autonomy in the Cardboard Publishing Movement' (*Journal of Latin American Cultural Studies*, 2016) and 'Viscous Porosity: Interactions between Human and Environment in Juan Rulfo's *El llano en llamas*' (*Journal of Iberian and Latin American Research*, 2015). Her current research project, on waste and waste pickers in contemporary Latin American cultural production, explores the lively interactions between human and other-than-human actants in 'wasted' environments and the ethical implications of these interactions in the context of the Global South.

Hannes Bergthaller is a Professor in the Department of Foreign Languages and Literatures at National Chung-Hsing University in Taichung, Taiwan. From 2012 to 2014, he served as president of the European Association for the Study of Literature, Culture, and the Environment (EASLCE). He is a research fellow of the Alexander-von-Humboldt Foundation and a recipient of the 2015 Wu Ta-You Memorial Award of Taiwan's Ministry of Science and Technology. Among his recent publications are a guest-edited cluster on ecocriticism and environmental history in *ISLE: Interdisciplinary Studies in Literature and Environment* (2015) and a guest-edited theme section on ecocriticism and comparative literature in *Komparatistik* (2014). His research focuses on ecocriticism, environmental philosophy, social systems theory and the cultural history of environmentalism in the US. His essays have appeared in *Connotations, Environmental Humanities,* the *European Journal of English Studies, English Studies, ISLE* and *New German Critique,* among other venues.

Claire Colebrook is Edwin Erle Sparks Professor of English in the English Department at Pennsylvania State University. Her many books include *Ethics and Representation* (1999), *Irony in the Work of Philosophy* (2002), *Deleuze and the Meaning of Life* (2010) and, with Tom Cohen and J. Hillis Miller, *Theory and the Disappearing Future* (2010). With Tom Cohen she is series editor for the Open Humanities Press series, Critical Climate Change.

Matthew Griffiths completed a PhD at Durham University in 2013. He has published a collection of poetry, *Natural Economy* (2016), and a monograph, *The New Poetics of Climate Change* (2017).

Adeline Johns-Putra is Reader in English Literature at the University of Surrey. Her books include *Landscape: Process and Text* (co-edited with Catherine Brace, 2009) and *The History of the Epic* (2005). Her essays on climate change and literature have appeared in *English Studies, WIREs Climate Change* and *ISLE: Interdisciplinary Studies in Literature and Environment.* She was chair of the Association for the Study of Literature and Environment UK and Ireland (ASLE UKI) from 2011 to 2015.

Richard Marggraf Turley is author of several books on the Romantic poets, including *Keats's Boyish Imagination* (2004) and *Bright Stars: John Keats, Barry Cornwall and Romantic Literary Culture* (2009). He is co-author with Jayne Archer and Howard Thomas of *Food and the Literary Imagination* (2014). Richard won the Keats-Shelley Prize for poetry in 2007 and recently published a novel set in 1810, *The Cunning House* (2015). He is Aberystwyth University's Professor of Engagement with the Public Imagination.

Chris Pak is editor of the Science Fiction Research Association's *SFRA Review* and the author of *Terraforming: Ecopolitical Transformations and Environmentalism in Science Fiction* (2016). For more information and links to articles and reviews, visit his website at http://chrispak.wix.com/chrispak.

John Parham was Professor of Environmental Humanities at the University of Worcester. He authored and edited seven books, including *Green*

Man Hopkins: Poetry and the Victorian Ecological Imagination (2010), *Green Media and Popular Culture: An Introduction* (2016), and most recently, *The Literature and Politics of the Environment: Essays and Studies* (2023). He co-edited the Routledge journal *Green Letters: Studies in Ecocriticism* for twenty years and published extensively on Victorian literature and ecology, green popular culture and contemporary environmental literature.

Dana Phillips is Professor of English at Towson University in Maryland and Senior Research Associate at Rhodes University in Grahamstown, South Africa. He is the author of *The Truth of Ecology: Nature, Culture, and Literature in America* (2003), and of articles, book chapters and reviews on a variety of ecocritical topics. His interests include the history of ecological theory and research, wildlife biology, the new materialism, and narratives of climate change and environmental apocalypse.

Kate Rigby, FAHA, is Professor of Environmental Humanities at Bath Spa University and Adjunct Professor at Monash University. Her research lies at the intersection of environmental literary, philosophical, historical and religious studies, with a specialist interest in European Romanticism, ecopoetics and eco-catastrophe. She is senior editor of the journal *Philosophy Activism Nature*, and her books include *Topographies of the Sacred: The Poetics of Place in European Romanticism* (2004), *Ecocritical Theory: New European Approaches* (co-edited, 2011) and *Dancing with Disaster: Environmental Histories, Narratives, and Ethics for Perilous Times* (2015). Kate was the inaugural president of the Association for the Study of Literature, Environment and Culture (Australia-New Zealand) and the founding director of the Australia-Pacific Forum on Religion and Ecology. She is currently a key researcher with the Humanities for the Environment Mellon Australia-Pacific Observatory.

Joshua Schuster is Associate Professor of English at Western University. His first book is *The Ecology of Modernism: American Environments and Avant-Garde Poetics* (2015). He is currently working on a new book: *What Is Extinction?*

Louise Squire completed her PhD at the University of Surrey in 2014. She also has an MA in Philosophy (Nature Pathway). Her research crosses the fields of contemporary literature and ecocriticism. Her doctoral work identified a theme of ecological death-facing in contemporary environmental crisis fiction. She has published several articles, including in the *Oxford Literary Review* (2012), and is working on her first book: *The Environmental Crisis Novel* (forthcoming). Louise is currently treasurer for the Association for the Study of Literature and Environment, UK and Ireland (ASLE-UKI) and assistant editor for *Ecozon@*.

Howard Thomas is Emeritus Professor of Biological, Environmental and Rural Sciences at Aberystwyth University and a Fellow of the Learned Society of Wales. He has held visiting professorships at universities in Switzerland and the US. His research interests include the genetics, evolution and uses of food plants. He also has a special interest in the cultural significance of scientific research and the promotion of links between science and the arts. He is co-author of *The Molecular Life of Plants* (2013) and *Food and the Literary Imagination* (2014). His most recent book is *Senescence* (2016).

Gillen D'Arcy Wood is the Langan Professorial Scholar of Environmental Humanities at the University of Illinois, Urbana-Champaign. His research takes the form of revisionist environmental histories of the global nineteenth century, with a special focus on climate change. The most recent of his four books, *Tambora: The Eruption that Changed the World* (2014), received worldwide attention – from *Nature,* the *New York Times* and *The Economist,* to the *South China Morning Post* and *Japan Globe* – and was recognised in the Book of the Year awards by the *Guardian,* the London *Times,* and the American Society for Literature, Science, and the Arts. His new book-in-progress, *Antarctica Through the Looking Glass: The Victorian Discovery Voyages, 1838–42,* reconstructs the vital role of the first South Polar expeditions in the origin of modern climate science and glaciology.

Foreword by Gillen D'Arcy Wood

The dirty secret of this volume, titled *Literature and Sustainability*, is that *there is no such thing as sustainability*. Extinction happens, sooner or later. And the pathos of that foreknowledge is what we call literature (among other names). A literary text is not self-aware, but it does have survival for its object: the words, in their making, intuit the chanciness of their conception, the horizon of their extinction, and seek out a niche of difference (not originality, which is why good books most resemble other good books). Literature is not death-driven, but rather auto-poetic, or ecological. The *literary*, though not itself living, enacts life, its own coming-into-being, together with the sense of an ending. Unlike other verbal artefacts then – memos, menus or internet ramblings – literature, in this self-sustaining sense, neither represents nor expresses. It is language itself hell-bent on survival; it rages against the dying of the light. A successful poem is no ordinary text; it is more akin to what ecologists call an 'idea model': a playful, data-light adumbration that captures essential features of a complex, real-world system, be it the psychology of love, a field of daffodils, or the death of kings (note to creative writing majors: if your poem doesn't have this capturing quality, it's unlikely to survive its first reading).

What does it mean to say literature is ecological? Mere words in a group do not become art – a thing we feel under a compulsion to revisit – unless they play with chaos, with the happenstance of their own being. Literary fictions, poems or prose, mimic the contingency and chaos of the bio-physical world, like Keats's 'salt sand wave' crashing, one time only, on the beach. Our own biophilia – one true gift of consciousness – is the source of our fascination with literature, including that which makes no reference to 'nature'. The literary, like organic forms in a complex system, is neither inevitable nor predictable. You read the poem over and it is never the same. You read it over and over because it is like falling in love with the true, contingent world, not some ideologised plastic replica you've been sold as happiness.

Complexity, for the ecologist, describes a self-organising system, based on simple rules and prone to chaotic behaviour. Such is literariness, that pleasure in texts late twentieth-century scholars were so eager to deny themselves. But the years of plenty are behind us. Scarcity looms. It's time to shore up the fragments of our ruins, to reconsider why literature, in the age of Twitter and downsized Humanities, might yet be indispensable to the Long Emergency of the twenty-first century. Complexity – which I am calling literariness – is not a first principle, but a material happening. It can happen in a frog pond; it can happen in the act of reading. Complexity derives from *plectere*, to weave. Penelope is the first and emblematic narrator of *The Odyssey*; she undoes and begins again on the same story with an ending she cannot fill in or foretell because she is living it. Complexity is a measure not of words or things but the amount of information a system stores. An Elizabethan sonnet is fourteen lines, but its commentary fills volumes. Complexity is an ecological measure, and the measure of literariness in a text. Far from being a disposable luxury item, literature – its body and being – is indispensable to sustainability as a concept because it's among the few examples we have of true resilience.

Sustainability, like survival, is a positive concept. It *happens*. It's ongoing. It *takes work*. Sustainability is thus best written in the imperative mood. In a 1923 poem, Robert Frost called it 'The Need of Being Versed in Country Things':

> The house had gone to bring again
> To the midnight sky a sunset glow.
> Now the chimney was all of the house that stood,
> Like a pistil after the petals go. (Frost 1995: l.1–4)

A farmhouse has burned down. Only the barn opposite remains. Over time, birdlife and plantlife reclaim the evacuated space. 'The birds that came to it through the air / At broken windows flew out and in' (13–14). Frost's poem is a parody of pastoral elegy, à la Gray's 'Churchyard.' In this world without us, neither the unsentimental farm folk who have abandoned their New Hampshire property, presumably for economic reasons, nor the poet-elegist sighing over 'what has been,' are validated. The poem's third way is ecological – an opening for the reader to see the complex intersection of human and natural systems by this 'stony road' that once belonged to the Great American Granary but no longer does, now repurposed for other species' flourishing: 'for them the lilac renewed its leaf' (17).

Frost's poem also usefully enacts our twenty-first-century predicament as humanists, and literary critics in particular. The poem is a mental struggle, for both poet and reader, a hard-won, incomplete transition from the safe haven of Romantic irony (weeping 'phoebes') to a post-humanist, ecological world view in which 'there was nothing really sad.' Frost's tone is the cue. The disaster that opens the poem is never signalled as tragedy, nor is the historical de-population of American rural places which is its crucial context. With the physical farmhouse gone, there remains only mindful, imaginative 'dwelling'. But this risks too much morbid brooding – 'the sigh we sigh / From too much dwelling on what has been' (15–16) – at the expense of observation of the non-human world. This the poem acts for us, as a caution against anthropocentrism, a lesson in sustainable perspective, a verbal 'idea model' – simple in design, brief in content – that captures essential characteristics of the infinitely complex human–nature dialectic over time. The birds, the flowers, the building and the elegiac eye are there, but so are the horse teams, farmers and fence-builders of time past:

> No more it opened with all one end
> For teams that came by the stony road
> To drum on the floor with scurrying hoofs
> And brush the mow with summer load. (9–12)

For all its impressive romanticisms of memory, the dark thought underlying Frost's poem is that the world does not care for us – a panic-room revelation very much of our time. The narcissisms of both reader and poet are rudely shattered by the indifferent birds and trees. Frost's cheerful risk here (pushing back against all that weeping and sighing) is also to gesture the consignment of his own poem to the scrap heap, like graffiti on an abandoned barn wall. But the wisdom of 'country things', and the poem that promotes them, teach us otherwise. Frost's 1923 poem survives, along with its birds and trees, because it is not the poem it was. It has been undone and rewoven, for retelling in a new century. What is left for us, the activist-readers of the twenty-first century, is to find a new and different niche for the human along the post-human lines the poem lays out. A niche, *not* a grave.

Why the urgent 'need' for rural knowledge? Frost's chaste poem never resorts to the tavern ribaldry of Shakespeare's doomed prince, teasing Ophelia about 'country matters' (*cunt-ry* matters). But Frost's demand that we know 'country *things*' has an erotic call all its own. The promised knowledge – 'being versed' – is itself *verse*: it must take on a beautiful shape to be sustainable. After all, without art and desire, nothing really lasts. Ideas are forgotten, feelings die and knowledge is lost. Although there is 'really nothing sad' in endings, as such, to be a literary survivor – to be 'versed' – is to understand the always tenuous resilience of natural forms, be it birdlife, or poems about birds, or the global human community transitioning (too slowly) to a post-carbon age.

The challenge of Frost's poem is to view his instructional scene – the abandoned barn – without irony. As the 'Foreword' contributor to this terrific volume of essays for our time, I challenge the reader to a thought-experiment. Sustainability without irony – is it possible? Can we put aside for a moment the cool lens of critique, together with the doomsday of sighs, weeping, and sadness that is the critical ironist's *lingua franca*?

Irony and critique – those Soho-style thought lofts of the metropolitan late twentieth century – appear to the extreme weather world of the late 2010s more and more like artefacts of the Anthropocene, like rhetorical gated communities for tenured humanists afraid to engage the holy mess we call the future. To escape these mind-forged manacles of the Late Petroleum Era, perhaps the word sustainability itself must go, to be dissolved into its purer form: *survival*. Survival – the concept – might serve for a sustainability freed of the taint of technocracy and corporate appropriation. And what better model of survival than literature, which enacts the resilience of all life worth the name?

Rising seas; the decimation (literally) of plant and animal life; the drain of life-giving aquifers ... Irony and critique are not the answer to these calamities. Knowledge and work are. (Remember Professor Serebryakov in *Uncle Vanya*: 'One must work!'). For all of us in the academic humanities, the pressing 'need' must be to explore well beyond our disciplinary comfort zones, to get ourselves 'versed in country things', be it the global climate system, the life cycle of the gannet or the rice futures market. For all of us, the need must be to better educate students who have never properly looked at a night sky, who have no idea who made their jeans, or how the animal died which fed them today. Knowledge and work: that's the unapologetic positivism of sustainability. For the rest, as for Frost's poet at the abandoned New Hampshire barn, everything not sustainability is just elegiac noise.

References

Frost, Robert 1995. *Collected Poems, Prose, and Plays*. New York: The Library of America.

Acknowledgements

The editors are deeply grateful to all who have supported or contributed to the process of bringing this book to fruition. We first wish to thank Matthew Jarvis, Anthony Dyson Fellow at the University of Wales Trinity Saint David, who was part of this book's inception and initially one of its editors, but who made the decision to stand back. We thank Matthew for his invaluable work on the initial stages of the book, and for his involvement in activities supporting initial research into literature and sustainability.[1] Our thanks also go to Jane Davidson, director of INSPIRE, University of Wales Trinity Saint David, for her part in these activities.

We wish to thank those who have provided valuable opportunities to present the research behind chapters in this book. We thank the organisers – especially Catriona Sandilands, York University, Toronto – of the Association for the Study of Literature and Environment (ASLE) eleventh Biennial Conference, hosted by the University of Idaho, at which we convened a panel on Literature and Sustainability. We thank the University of Surrey for funding one of our editorial team, Adeline Johns-Putra, to attend and present on the panel. We thank our panel chair Dana Phillips and contributors Chris Pak, Joshua Schuster and Louise Squire for supporting the panel. By the same token, we thank Páraic Finnerty and Mark Frost, University of Portsmouth, for allowing another of our editorial team, John Parham, to present a version of his chapter for the Centre for Studies in Literature (CSL)'s seminar series.

Our grateful thanks also go to all who have been directly involved in bringing this book together. We thank the staff at Manchester University Press for their helpful guidance and support. We thank Gillen D'Arcy Wood for providing the book's foreword. We thank Lucy Bell for kindly allowing us to use an image of her Cartoneras collection, acquired through her research activities, for the book's front cover. We thank photographer Vassilis Korkas, who took the picture on Lucy's behalf. On behalf of Lucy, we also thank Lucia Rosa and the Dulcinéia Catadora cooperative for their time, support, materials and inspiration. Finally, yet foremost, our thanks go to all our contributors, whose excellent essays come together to make this book the ground-breaking venture we hope you find it to be.

Notes

1 These include an Association for the Study of Literature and Environment, UK and Ireland (ASLE-UKI) symposium, which was supported financially by Institute of Sustainable Practice, Innovation and Resource Effectiveness (INSPIRE) and hosted by the University of Wales Trinity Saint David; a public lecture competition organised by Jane Davidson, director of INSPIRE, and supported by ASLE-UKI; and a special issue of *Green Letters: Studies in Ecocriticism* (19.1), for which Matthew was co-guest-editor with Louise Squire.

Editors' introduction

Adeline Johns-Putra, John Parham and Louise Squire

The overall aim of this collection is to explore the ways in which literary scholarship might engage with and contribute to the sustainability debate. 'Sustainability' *per se* has been slow to acquire interest as a concept for literary scrutiny, despite its ubiquity in the cultural and socio-political present, and despite the ambitious range of work emerging in the relatively new field of ecocriticism. Even so, it has not been altogether absent from literary scholarship, as indicated, for example, by references in this book to several essays in the theories and methodologies section, entitled 'Sustainability', of the May 2012 issue of *PMLA* (*Publications of the Modern Languages Association of America*), and by the more recent January 2015 special issue – 'Literature and Sustainability' – of the journal *Green Letters: Studies in Ecocriticism.*[1] 'Sustainability' of course has its problems, being a fraught, paradoxical and contested term with a spectrum of definitions, applications and uses. Nonetheless, our purpose in bringing together this collection is not to seek to overcome sustainability's difficulties. Nor is it to redefine it or to view it from any particular stance. Rather, it is to demonstrate, through the essays presented, the various ways in which literary scholarship might reflectively engage with and comment upon sustainability and, in doing so, to illustrate what an engagement with sustainability might offer to literary and ecocritical scholarship. In this introductory section we offer some thoughts on sustainability and its difficulties, discuss its employment as a critical concept and consider it as a question for literary scholarship. We also provide an overview of the book, to include an outline of its two main sections and summaries of each of the essays.

Difficulties arising in sustainability discourse

A key premise for this book is that it is not just despite but because of its difficulties that a literary engagement with sustainability can prove variously productive. Some of these difficulties revolve around its adoption as a mainstream response to environmental concerns. While sustainability might be variously historicised (see the first two essays in this collection – by

Jayne Elisabeth Archer, Howard Thomas and Richard Marggraf Turley, and by John Parham), it is of course strongly influenced in current parlance by its most prolific definition, as derived from the Brundtland Report. This report places emphasis on humanity's long-term survival through a notion of 'sustainable development', which it describes as development 'that meets the needs of the present without compromising the ability of future generations to meet their own needs' (World Commission 1987: 43). Such a definition, in setting a concrete goal applicable across domains, has given rise not just to sustainability's widespread implementation but also to the extensive body of work that continues to redefine sustainability today.

Since ecocritics have often been most interested in approaches that strive, in literary and philosophical terms, to reach beyond a reductive human-centrism, sustainability's emphasis on the human sphere has led to concerns that it fails to consider the nonhuman world sufficiently, its agencies and our kinship with it (Alaimo 2012; O'Grady 2003). Its instrumentalist demeanour can bring it into conflict with the deeper green standing of many ecocritics (Bergthaller 2010; O'Grady 2003). A further difficulty is that, due in part to sustainability's slipperiness as a term, it is vulnerable to co-option (Alaimo 2012; Bloomfield 2015; LeMenager and Foote 2012; Nardizzi 2013; Squire and Jarvis 2015), whereby it is appropriated to legitimise corporate or political endeavour – to include, as Stephanie LeMenager and Stephanie Foote put it, that which 'got us into trouble in the first place' (2012: 572; see also Nardizzi 2013). Such problems, as Lynn Keller observes, may come down to a question of 'sustainability of what and for whom?' (2012: 579).

Sustainability also raises a number of conceptual difficulties. One worry is that it erroneously strives for some kind of stasis whereby the ecological present or past is extended into the ecological future (Bloomfield 2015; O'Grady 2003; Mentz 2012). John P. O'Grady, for example, sees an 'obvious flaw in reasoning' in sustainability's 'privileging of *duration* or *permanence* as a value', which 'runs counter to' the 'fundamental principle' that nature 'is in flux' (2003: 3, emphasis in original). Mandy Bloomfield similarly views sustainability as a 'comforting narrative' in need of 'unsettling', and points out that the sciences of ecology 'have generally moved away from equilibrium-seeking ecosystems towards disequilibrium models' (2015: 4–5). On this note, Gillen D'Arcy Wood emphasises the need for ecocritics to recognise the 'emergent biocomplexity paradigm driving sustainability science' (2012: 8), observing that sustainability studies 'begins from the principle that all systems, human and natural, are characterised by complexity and nonlinear change' (6).

Other conceptual difficulties include its various internal conflicts. Adeline Johns-Putra, in this collection, remarks for example on the way sustainability appears as an exhortation to both preserve and transform. That is, the project of safeguarding our own future by sustaining planetary wellbeing is reliant upon our capacity and willingness to enact effective change. Yet, as she remarks elsewhere, while 'our construction of "sustainability" is driven by a notion of care' for the nonhuman world, care is itself a variously contingent concept that lacks 'ontological scrutiny' (2013: 125). In engaging with futural notions of our own and other species, sustainability infuses the present with that which is yet to emerge, and the subject of the present with that which necessarily exceeds it. Effects of this difficulty include the way objectives such as sustaining species diversity take on a managerial approach to that which managerialism also undermines; sustainability's vulnerability to co-option is perhaps another outcome. Yet, as essays in this collection also indicate, such conceptual challenges are themselves some of the grounds upon which sustainability might be explored from a literary and ecocritical perspective.

Towards a critical sustainability

If ecocritics have reason to resist sustainability's nebulous and at times contentious constructed vision(s), some contra-points might also be made which begin to move us towards the goals of this book. Sustainability's difficulties might be said to open up a productive opportunity for interrogation and exploration of the kind that literary scholars are ideally placed to carry out. Indeed, critics such as Hildegard Kurt or LeMenager and Foote have referred to a 'cultural deficit' within the sustainability concept, recognising that 'academic humanists and artists have not been central to discussions of what sustainability is and might be'; consequently they call for a sustainable humanities (LeMenager and Foote 2012: 572). After all, if practices of sustainability have infiltrated the socio-cultural, political and correspondingly the literary present, a question for environmental literary scholarship has clearly emerged.

So why critical sustainability, and what do we mean by it? There is a politically critical tradition of sustainability which, in the work of Fischer et al. (2007: 622), reshapes the so-called three pillars of sustainability – the economic, social and biophysical – into a more hierarchical paradigm, one where ecological sustainability envelops social sustainability. Indeed, Kate Rigby's essay, in this collection, reflects on Fischer et al.'s work by considering literature's potential contribution to 'this cultural work of "deep sustainability"'. Here, however, we mean a range of responses to sustainability from within the field of environmental literary criticism that might contribute to sustainability's broader debates. That is, the phrase describes a literary response to sustainability variously explored, as demonstrated in the differing approaches taken in this collection.

Within ecocriticism today a question appears as to whether, and how far, ecocritics should engage in political polemic through literary practice. More activism or less activism? Much, although certainly not all, of the discomfort with sustainability referred to above could be described as broadly political in character, since it represents a more radical resistance, not necessarily to sustainability *per se* but certainly to some of its mainstream applications. One of the premises upon which this book rests is that sustainability's adoption in a political mainstream need not render it obsolete to ecocritical inquiry. This is specifically because we see the role of contributing to the sustainability debate as a critical (that is, a literary critical) role.

Indeed, the more activist ecocritic might choose to get involved in (re) conceptions of the term based on literary innovation, with a view to considering how sustainability might be redeemed as a concept for literary inquiry. For example, a critical sustainability could reposition the term towards a reflection on who we are in relation to others (i.e. the insight offered by posthumanism) – and our vulnerability in that condition – rather than asking us to care in ways that are abstract and hard to convince. Or it might be premised upon a negotiation between two recognitions – that, at some deep level, human existence remains answerable to the realities of biophysical support systems, but also that ecosystems exist in a state of flux and discordance within which humans have considerable (though not absolute) latitude to co-construct the nature in which we live; thus it would engender moral, ethical, social choices as to what type of 'sustainable society' we might construct. Accordingly, rather than aspiring towards a utopian 'stationary state', sustainability might be seen as something the human species has continually to strive towards, while knowing it will never reach it.

The aim of this book, however, is not to redeem sustainability as a normative concept but to view it, less prescriptively, as a concept open to exploration and debate, and as potentially opening up a space for new innovations in environmental literary scholarship, and correspondingly, recognising where and how literature probes the thorny question of what it might mean to live sustainably. Certainly, these seem to be opportunities that sustainability itself provides, being fraught not just with conflict but with paradox. Hannes Bergthaller, in this collection, argues that it is precisely because of the dialectic of sustainability's competing priorities, which require constant renegotiation, that sustainability constitutes 'genuinely political matter'. On this view, approaching sustainability from an environmental literary perspective retains a political dimension, but one that is explorative and reflective. Arguably, such reflective exploration might be considered a deeper political praxis than attempting to prise sustainability into any particular shape or form.

Ultimately, sustainability encompasses too much to reduce it to singular statements. So, for all its difficulties, sustainability also shifts debates in ways that might actually enhance and add to established ecocritical discourse. As Simon Dresner notes, sustainability is 'much more powerful rhetorically' than the term 'environmentally friendly'; since 'publicly saying that you don't care that what you are doing is unsustainable sounds tantamount to admitting that you are intellectually incoherent' (2002: 1). For Dresner, sustainability may be a 'contestable concept' (2002: 7), but – like other such slippery terms as liberty and justice – it is useful as a discursive starting point. He also reminds us that the sustainability agenda introduced by the Brundtland Commission represents the first time the question of equity within generations had been balanced with the question of equity between generations. Moreover, its acceptability by the mainstream opens up possibilities of exploring the mainstream imagination. Whatever its weaknesses, then, discourses of sustainability run in tandem with some very pressing – and very present – socio-political and philosophical conundrums: literary scholarship that engages with sustainability is therefore engaging with such key questions. 'Critical sustainability', as we frame it in this volume, thus points simply towards a critically reflective approach to the problem of sustainability – an approach that we argue is not just timely but urgent. Such a role is, as much as anything, a literary role, whereby sustainability's difficulties and possibilities might be teased out and explored.

Literature and sustainability

In discussing sustainability from a literary perspective, we draw forward two approaches that broadly correspond to those demonstrated by the essays in this volume. One, as indicated in the previous section, is that of a critical sustainability. Certainly, other literary scholars have suggested that the very discourse and praxis of sustainability bears scrutiny of a literary kind. Karen Pinkus has argued that sustainability functions in the same way as narrative; it 'implies or writes a narrative coherence' (2011: 74), and rethinking sustainability requires that we rethink narrative itself. Indeed, a narrative of *jouissance* rather than of futurity might release us from the trap of 'business-as-usual' thinking that accompanies so much sustainability discourse. The other approach may be considered a literary response (broadly speaking) to such discourses of sustainability, including an emphasis on the possibilities that arise in a fluid engagement with literature *per se*. LeMenager and Foote argue that 'the most complex and wide-ranging intersection between literary studies and sustainability lies at the intersection of literary forms and social affiliation', that is, 'in how literary forms prompt us to imagine, as communities, a world otherwise' (2012: 575). Similarly, we see literature, literary form and literary scholarship as contributing a way of engaging with sustainability in an imaginative sense. Lynn Keller views literature as helping us to imagine the risks current behaviour is running, as potentially evoking an 'almost unimaginable' large-scale, long-term perspective on sustainability, or as engaging us, whether through narrative or poetic language, in 'serious imaginative exploration of what constitutes a desirable future' (2012: 581–2). Keller regards this as a 'literature not of but pointing toward sustainability' (2012: 582). The imaginative dimension that Keller proposes suggests a way of exploring literature and sustainability that avoids being overly prescriptive; for, in its fit with the discursive nature of the sustainable, literature poses complex conceptual questions and models of how we live with the other forms of nature on which we depend, or imagines what types of societies might be sustainable. Literature might equally, we would argue, provide a space in which to explore the complexity of sustainability as an ongoing, never fulfilled aspiration, or the difficulties of attaining a sustainable world, or the nuances and dimensions of the unsustainable. Indeed, Claire Colebrook, in this collection, illustrates how the 'logic of literary sustainability' might lead us to question the value of the future of 'what has inscribed itself as humanity'. In practice, to consider what insights literature might offer sustainability debates also requires a close consideration – a close reading, in many cases – of literary strategies, genres and theory, as well as literary history.

Part I of the collection – 'Discourses of sustainability' – presents six essays that approach sustainability in a variety of ways. It begins with studies of literary texts in their historical contexts. In their investigation of literary mills, Jayne Elisabeth Archer et al. show how these are not mere aesthetic representations but indicators of sustainable practice in their own times. John Parham's chapter then demonstrates how nineteenth-century novelists William Morris and Emile Zola were attuned to – and keen to express in their work – ideas in Britain and France respectively that we would now identify as anticipating sustainability. As other essays in Part I amply demonstrate, one can regard both literature itself and literary scholarship as acts offering insights on, working with, and indeed re-working the discourses of sustainability that shape how effectively (or not) our individual and societal practices facilitate the continuation of ourselves and the other species with whom we co-exist. Kate Rigby discusses a model of sustainability that she labels 'deep sustainability'; reading retrospectively, she finds expressions of this model in the poetry of John Clare and practice of the contemporary writer David Morley. Lucy Bell and Joshua Schuster then proceed in a spirit of discursive inquiry. Bell subjects the much-vaunted 'three pillars' of sustainability to critique via an example not from literary text but from literary production. She discusses the operations of the Latin American *editoriales cartoneras* publishing houses, where books are handmade from recycled cardboard and often individually produced and sold, as a material exemplar of sustainable literary intervention. Schuster, meanwhile, interrogates a very different kind of sustainability practice – the conservation of small populations of nearly extinct species. He shows how such projects as the rewilding of bison in North America might be invested in naive notions of sustainability as return but, when examined closely, serve to challenge such assumptions. In the final essay in Part I, Claire Colebrook offers a theoretical perspective on the question of literary sustainability, arguing for the use of a deconstructive or material sublime in reading the Anthropocene at a time when thoughts of sustaining ourselves arise.

What then follows in Part II – 'Reading sustainability' – is a number of chapters that employ close readings as a lens through which to critique contemporary sustainability discourse. Dana Phillips demonstrates how a sensitivity to genre – in this instance, the historical novel – shows the flaws in sustainable thinking. While humans might strive for sustainability, the threat of collapse and the associated development of characteristics of resilience are what dominate the cycles of nonhuman nature; for Phillips, it pays to read Margaret Atwood's MaddAddam trilogy, which traces just such a cycle, as a historical novel of environmental behaviour and, simultaneously, as a critique of our commitment to sustainability. Chris Pak's exploration of science fiction's historical engagement with environmental issues next suggests ways in which literature might provide signposts to more effective modes of sustainability, showing how the genre has long attempted to provide sustainable solutions – an effort recently exemplified by Kim Stanley Robinson's Science in the Capital trilogy. Adeline Johns-Putra and Hannes Bergthaller then read the legacy of Brundtland through two contemporary novels. Johns-Putra's essay on Jeanette Winterson's 2007 novel, *The Stone Gods*, critiques the failure of deep green versions of sustainability to account for Brundtland's preservative – and one might say conservative – tendencies, and shows how Winterson's work constitutes a perhaps unwitting exposé of this failure. Bergthaller's analysis of Michel Houellebecq's *The Possibility of an Island* (2005) demonstrates how the sustainability agenda is premised on the mistaken assumption that ecological sustainability and emancipatory politics go hand in hand: this assumption is deconstructed to devastating effect in Houellebecq's dystopia. Matthew Griffiths reads sustainability as embedded within the very form of Jorie Graham's *Sea Change* – the effort to sustain one's breathing in reading Graham's poems aloud echoes the pains, both psychological and physical, that we take to live sustainable lives. Finally, Louise Squire's speculative realist account of Yann Martel's *Life of Pi* (2001) pays heed to the seemingly irresolvable tension between sustaining the human and preserving the nonhuman that lies at the crux of sustainability discourse: Martel's Pi, like all human subjects, must go beyond a merely correlationist or phenomenological engagement with the nonhuman world if he (and we) are to resolve this tension. Such literary readings show how close analyses of novels can productively trace the faultlines and fissures of sustainability discourse.

Ultimately, we argue, the fraught, paradoxical, contentious, yet ubiquitous nature of sustainability, as a concept in need of interrogation in our time, seems actively to call for the reflectivity of critique that literary scholarship is well placed to carry out even if, ultimately, all literature can do is offer signals and pointers. Indeed, a recent 'provocation' in Volume 5 of *Environmental Humanities*, the authors of which include several of the contributors to this collection, specifically calls for 'a greater reflexivity' within the environmental humanities, and one that might at times 'be at odds with' those discourses of environmentalism that otherwise resist such critique (Bergthaller et al 2014: 268). Sustainability, we suggest – and as we hope the essays in this collection amply illustrate – both invites and demands this kind of literary critical attention.

Notes

1 *Green Letters: Studies in Ecocriticism* is the journal of the Association for the Study of Literature and Environment, UK and Ireland (ASLE-UKI).

References

Alaimo, Stacy 2012. 'Sustainable This, Sustainable That: New Materialisms, Posthumanism, and Unknown Futures', *PMLA* 127 (3): 558–64.

Bergthaller, Hannes 2010. 'Housebreaking the Human Animal: Humanism and the Problem of Sustainability in Margaret Atwood's *Oryx and Crake* and *The Year of the Flood*', *English Studies* 91 (7): 728–43.

Bergthaller, Hannes et al. 2014. 'Mapping Common Ground: Ecocriticism, Environmental History, and the Environmental Humanities', *Environmental Humanities* 5: 261–7.

Bloomfield, Mandy 2015. 'Unsettling Sustainability: The Poetics of Discomfort', *Green Letters: Studies in Ecocriticism* 19 (1): 21–35.

Dresner, Simon 2002. *The Principles of Sustainability*. London: Earthscan.

Fischer, Joern et al. 2007. 'Mind the Sustainability Gap', *Trends in Ecology and Evolution* 22 (12): 621–4.

Johns-Putra, Adeline 2013 'Environmental Care Ethics: Notes Towards a New Materialist Critique', *Symplokē* 21 (1–2): 125–35.

Keller, Lynn 2012. 'Beyond Imagining, Imagining Beyond', *PMLA* 127 (3): 579–85.

LeMenager, Stephanie and Stephanie Foote 2012. 'The Sustainable Humanities', *PMLA* 127 (3): 572–8.

Mentz, Steve 2012. 'After Sustainability', *PMLA* 127 (3): 586–92.
Nardizzi, Vin 2013. 'Greener'. In *Prismatic Ecology: Ecotheory Beyond Green*. Ed. J. J. Cohen. Minneapolis: Minnesota University Press, 147–69.
O'Grady, John P. 2003. 'How Sustainable is the Idea of Sustainability?', *Interdisciplinary Studies in Literature and Environment (ISLE)* 10 (1): 1–10.
Pinkus, Karen 2011. 'The Risks of Sustainability'. In *Criticism, Crisis, and Contemporary Narrative: Textual Horizons in an Age of Global Risk*. Ed. Paul Crosthwaite. New York: Routledge, 62–77.
Squire, Louise and Matthew Jarvis 2015. 'Editorial: Literature and Sustainability', *Green Letters: Studies in Ecocriticism* 19 (1): 1–7.
Wood, Gillen D'Arcy 2012. 'What is Sustainability Studies?', *American Literary History* 24 (1): 1–15.
World Commission on Environment and Development 1987. *Our Common Future* (Brundtland Report). Oxford: Oxford University Press.

Part I

Discourses of sustainability

1

The millers' tales: sustainability, the arts and the watermill

Jayne Elisabeth Archer, Howard Thomas and Richard Marggraf Turley

In 2009, the Nobel Prize-winning economists Joseph E. Stiglitz and Amartya Sen issued a report urging a shift from a purely economic analysis of a country's success or relative failure to one which includes (and is informed by) an analysis of wellbeing and sustainability (Stiglitz et al. 2009). The report concluded that wellbeing and sustainability, which comprise factors such as culture, education, health, water security and food production, are intimately linked. Although their terminology and modes of communication may have differed, the artists and writers of the past have also been attuned to this connection – a connection many of us today have almost lost – and to the various pressures that have threatened to undo it.

The watermill in time

An important but often neglected site in the relationship between literature and the visual arts on the one hand, and sustainability on the other, is the watermill. Our concern here is with the water-driven mill, while we acknowledge there is also a tale to be told about windmills. Wind is a fickle source of power, as opponents of modern wind farms like to point out. Water, by contrast, is seemingly more controllable and predictable – in this sense, more sustainable – than wind, and therefore a more stable centrepiece of community life throughout the world and history. The watermill is frequently sentimentalised as what Terry S. Reynolds has called a 'picturesque artifact' in the modern mind, and abstracted from specific historical moments and social forces (Reynolds 1983: n. pag.). For many hundreds of years the watermill was the point at which food entered most transparently and immediately into the worlds of politics, governance, culture and social justice. It was a complex site within which communities were created and negotiated, through cultural as well as material relationships. The importance and intricacy of the work performed in and by the watermill elevated it to symbol, ritual, myth and mystery. But, for as long as it remained an everyday part of town and village life, it was also an insistent and shaping material presence. The skilled miller, sifting through the grain, was an important guard against the corruption of the food chain by toxic agents such as darnel and ergot (Archer et al. 2014). The watermill made it possible for the owners of smallholdings to work their land and feed themselves. In it, the ongoing conflict between country and city, and their very different appetites, was played out. It was a place of urgency and contention, in which weights and measures – customary and standardised – were debated and resisted. In short, the watermill had an essential role in the formation of pre- and early modern communities: it enabled them to be self-sustaining; it made the people and their land sustainable.

There are several rich accounts of the history of the watermill in Britain – for example by Reynolds (1983), Steven S. Kaplan (1984) and Martin Watts (2006) – and Beryl Rowland (1969, 1970) has surveyed literary (including classical) representations of milling and millstones. Reynolds suggests that its very ubiquity in history and literature has made the watermill an overlooked subject for contemporary cultural and ecocritical study (1983: 3). This desertion perhaps also results from the fact that for much of British history (and English literary history), the physical structure of the watermill itself appeared resistant to change: in the period 1300–1850, the basic machinery and processes used in the watermill remained much the same (Reynolds 1983: 3). We see the unquestioning acceptance of traditional custom and practice in the lack of a definitive answer to, or even curiosity about, whether and to what degree overshot waterwheels deliver more power than undershot. It was not until 1759 that the engineer John Smeaton (builder of the Eddystone Lighthouse) finally resolved the matter in a paper to the Royal Society. As a result of experimental and mathematical modelling – among the earliest examples of the application of scientific method to engineering – Smeaton showed conclusively that the overshot wheel is twice as efficient as the corresponding undershot wheel (Capecchi 2013).

It's a familiar story: scientific insight and technological advance that lead to first gradual, then rapid, sweeping away of 'inefficient' tradition and, with it, of hitherto homeostatic communities and cultures. We find that, just as much as the surrender of common ground to successive waves of enclosures, the loss of the watermill as a centre of food production – owned and operated by and for the community – marks a fault line, a profound trauma in British history. Industrialisation replaced the grain mill with the mills of manufacture – cotton, paper, wool, steel, as elegised by Richard Jefferies and celebrated by J. M. W. Turner (Jefferies 1880; Rodner 1997). The mill is a recurrent mystical symbol in the writings of William Blake and even has a walk-on part in the early history of the information age: the analytical engine of Charles Babbage comprised the 'store' and the 'mill', precursors of the memory and central processor of modern computers (Swade 2002: 105).[1]

One of the most famous literary watermills instructed its cultured, largely urban readership in the dangers of neglecting – and, importantly, neglecting by misreading – the watermill as a site. Although *Don Quixote* (1606, 1615) is better known for its 'tilting at windmills' episode, Cervantes' antique knight makes a similar mistake when he approaches two watermills. Sancho Panza, a former farmer, sees what is before him: 'two large watermills in the middle of the river', which he further explains as 'watermills ... where they grind wheat'. Don Quixote sees something quite different. '[A]lthough they seem to be watermills', he explains, 'they are not': 'There, my friend, you can see the city, castle, or fortress where some knight is being held captive, or some queen, princess, or noblewoman ill-treated, and I have been brought here to deliver them' (Cervantes 2005: 650)

Their boat caught in the fast-flowing millrace, it is Quixote and Sancho who have to be rescued by two floury-faced and exasperated millers. As Harry Levin remarks, in the figure of Don Quixote Cervantes explores the relationship between 'literary artifice and that real thing which is life itself' (Levin 1959: 81). Elsewhere, we have considered the tendency among scholars and literary critics to read literary representations of watermills as something, anything, other than what they are, and for what they do: namely, places 'where they grind wheat' (Archer et al. 2015a). Like the Golden Age knight, on occasions literary critics and modern readers should perhaps be willing to attend to the words of Bishop Joseph Butler: 'Everything is what it is, and not another thing' (1726: 19). Watermills are 'real things', places where wheat is ground. Watermills happen to people, and as a result they are material presences in literature and culture. The cost of neglecting to consider the watermill in such terms is to fail to understand the important lessons millers' tales can tell us about the role of food production in sustainability – lessons that are vital to our own and future wellbeing.

Tales of water, wheat and self-sustaining communities

In poetry, prose and the visual arts, Britain has been celebrated as an island formed and powered by the interaction of water and wheat. In the frontispiece to William Camden's *Britannia* (1586, 1610: 3), a cartographic rendering of the British Isles, seemingly certain in its locations and relative dimensions, is accompanied by two classical deities (Figure 1.1). To the left is Neptune, god of the sea, and to the right is Ceres, the corn goddess. Just as much as the mapped part of this frontispiece, the presence of the

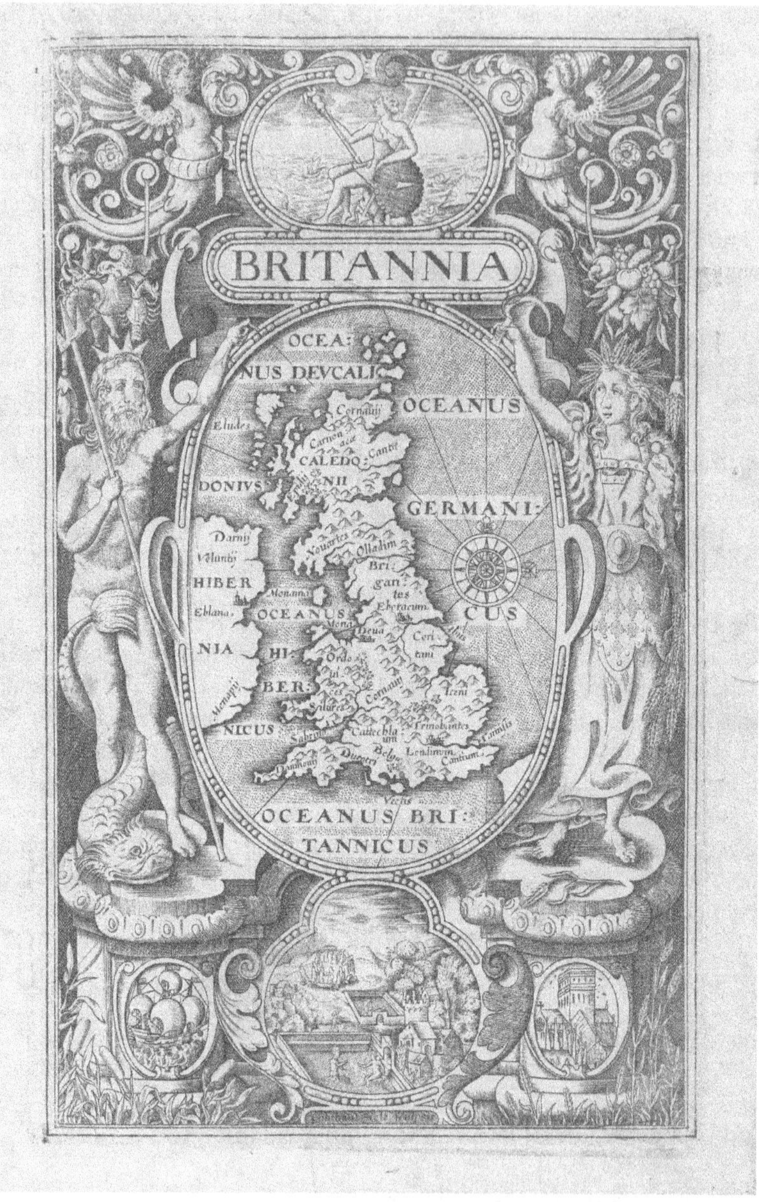

Figure 1.1 Illustrated frontispiece, William Camden. 1610. *Britannia, or A Chorographicall Description of the Most Flourishing Kingdomes, England, Scotland, and Ireland, and the Ilands Adioyning*

gods of sea and corn reveal an important truth about the history of these islands. The matter of Britain is enlivened by the meeting and interaction of sea and land, water and wheat – what Thomas Hobbes, writing in 1651 called 'the two breasts of our common Mother' (1985: 285). When these resources are exploited in a sustainable fashion, Camden explains, Britain can not only feed herself, she can afford to export overseas, thereby fuelling her own imperial ambition.[2]

Writing over 200 years later, John Keats was also able to imagine a 'Kingdom of Corn', albeit one no longer associated with a particular place in the present, but one seemingly lodged in the mythical past: the golden age of a Virgilian autumn. Keats's Apollo addresses the three Graces:

> Which of the fairest three
> To-day will ride with me?
> My steeds are all pawing at the threshold of the morn:
> Which of the fairest three
> To-day will ride with me
> Across the gold Autumn's whole Kingdom of corn? (Keats 1988: 56)

Embedded within these seemingly abstract and timeless visual and literary references is an urgent and determinedly time-bound politics of food supply. Camden and Keats wrote not in times of abundance, but in times of dearth. The period 1580–1610 witnessed a run of poor harvests. In a series of initiatives, the state attempted to control the production, processing and distribution of grain; when those measures were perceived to fail, riots broke out in London and the Midlands.[3]

Keats wrote amidst febrile debates concerning responses to the spiralling corn prices generated by the 1815 Corn Law and cheap labour exacerbated by the influx of soldiers returning from the Napoleonic Wars (Barnes 1930: 117–84; Gash 1978). Proposals for a second Corn Bill were debated in Parliament during late 1818 and early 1819. The impact of these factors on the prices and distribution of food led to increasing food insecurity (and profoundly influenced Keats's poetry at that time; see Marggraf Turley et al. 2012). It was a situation likely to result in revolution, as Byron warned in *The Age of Bronze*:

> For what were all these country patriots born?
> To hunt, and vote, and raise the price of corn?
> But corn, like every mortal thing, must fall,
> Kings, conquerors, and markets most of all. (Byron 1823: 28)

Allusions to corn, wheat and harvests in the works of Camden, Keats and Byron are not simply reworkings of a literary trope as old as Hesiod and Virgil. For all three British authors, the corn they write about is pressingly real and is part of a wider web of environmental conditions, political imperatives and socio-economic concerns – unsustainable times, with uncanny similarities to our own. Indeed, the interplay of water and wheat is, in turn, part of a much bigger story concerned with the sustainability of food production and distribution. Globally, more than 80 per cent of the land used for growing crops depends entirely on precipitation to support plant production. The remaining cropland is irrigated and supplies almost 40 per cent of the world's food and fibre needs. Ours is a thirsty planet. It takes about 500 tons of water to make 1 ton of potatoes. A ton of wheat needs 900 tons, maize 1,400 tons, and rice comes in at a mighty 2,200 tons (Mekonnen and Hoekstra 2010). The international trade in food can therefore be understood an international trade in water.

In the sciences, a proxy is a measurement of one physical quantity that is used as an indicator for the value of another. Food can be considered to be a proxy for both water and for the chemicals required for human nutrition. Former ages were attuned to this fact: for example, owners of watermills who diverted water for their own use were considered guilty of 'hoarding' water – a crime equivalent to hoarding grain or other staples (Kaplan 1984: 225). The food chain, then, can be seen as a proxy for the flow of carbon (or nitrogen, phosphorus and sulphur) from agriculture to consumer. Alternatively, it can be seen as a proxy for the transfer of solar energy trapped by photosynthesis from autotrophs (green plants) to heterotrophs (humans, animals and all the other plant-dependent life forms). It can even become a proxy for information (from the genotypes of plants to the genotypes of animals, mediated by their phenotypes). How we think of sustainability depends on the particular proxy we favour. Proxies are essentially metaphors. And metaphor is nothing if not the warp and weft of literature.

In literature, from medieval times to the dawn of the industrial era, the watermill performed (to quote Byron) an 'Agrarian Alchymy': a dynamic, evolving force in narrative and metaphor, it helped to trace the changing imperatives within, and pressures felt by, self-sustaining communities (Byron 1823: 28). In earlier literary and cultural representations, the watermill was usually accommodated within specific interpretive schemes. Classical myth imagined the world as a giant mill, with the gods as the millers who grind mortals through lifetimes of suffering and endurance (Rowland 1970: 215–16). Echoed in the Old Testament, this symbolism was carried over into the Christian tradition of the 'Mystic Mill,' in which threshing and grinding signified the apocalypse and Last Judgement, pure souls separated from impure. As the 'Mill of the Host' in medieval times, the watermill carried Eucharistic associations, representing the transformation by which Old Law is 'ground' into the New Law, and flour is transubstantiated into the body of Christ, the Host (Aston 1994; Delasanta 2002).

In the English literary canon, concerns about sexuality are never far from issues of class and legitimacy (sexual and parental). In post-medieval drama, such as John Fletcher and William Rowley's comedy *The Maid in the Mill* (licensed for performance in 1623 and first published in 1647) and the anonymous *Fair Em, the Miller's Daughter of Manchester* (performed between 1589 and 1593), the watermill is a metaphor for the purity (or otherwise) of sexuality and breeding, and serves both to conflate these anxieties and to project them onto the body of the miller's daughter. In *The Maid in the Mill*, the name of the miller's daughter, 'Florimel', carries echoes of both 'flour' and 'flower', alluding to her social rank as well as her virginal status. But, just as a miller has to be able to recognise weeds in order to produce the finest (and, for the consumer, safest) flour, a woman needs to know how to play the harlot in order to preserve her chastity, and Florimel pretends to be sexually experienced in order to escape her abductor, Count Otrante (Fletcher and Rowley 1909).

The Renaissance stage used milling and grinding as metaphors for sexual intercourse – especially intercourse of an illicit variety.[4] Although William Shakespeare made extensive use of this metaphor, it is significant that one of the two mentions of actual mills in his works is of a paper mill (in *Henry IV Part 2* [1597–8]: 4.7.67); the other allusion is made by Poor Tom as part of the ancient, broken world of *King Lear* [1605]: 3.1.90–3). Perhaps because of the playwright's extensive acquisition of food-producing land in Warwickshire (see Archer et al. 2015b) – but also, no doubt, to boost sales – the first printed edition of *Faire Em* was attributed to Shakespeare (Chambers 1923: iv.11; Tucker Brooke 1908). In this comedy, set during the reign of William the Conqueror, the miller is a disguised nobleman who regains his rank by the end of the play, and his chaste daughter, the eponymous Em, represents not simply her father's honour but that of the pre-Conquest English. First performed in the wake of the Spanish Armada, *Faire Em* celebrates the miller as representative of Englishness itself, and it is perhaps no coincidence that to Elizabethans, France was notorious for the poor quality of its cereal crops and bread.[5]

By the late eighteenth and early nineteenth centuries, literary portrayals of the watermill assume an elegiac tone, whether in the tragi-comedies of George Eliot's *The Mill on the Floss* (2010, first published 1860) and Thomas Hardy's *The Trumpet Major* (1880), or the maudlin poetry of Robert Bloomfield.[6] Alfred Lord Tennyson's 'The Miller's Daughter' (published in 1833 and again, with substantial revisions, in 1842) was supposedly inspired by Trumpington Mill in Cambridgeshire (Pinion 1984: 87), but this poem, like Bloomfield's, is abstracted by its author from time and place, the lost life or love of the miller's daughter standing in generalised terms for the lost communities and ways of life sustained by the watermill.

In the second half of this essay, we turn to the stories of four watermills, actual and (re)imagined: Flatford Mill in East Anglia, often associated with John Constable's 1821 painting now known as *The Hay Wain*;

Trumpington Mill in Cambridgeshire, setting for Geoffrey Chaucer's *The Reeve's Tale* (*c.*1390); Dorlcote Mill, home to the Tulliver family in George Eliot's *The Mill on the Floss*; and Felin Ganol in Llanrhystud, Ceredigion, a building with medieval origins and recently restored to working order. The stories of these watermills, the tales told by and about their millers, mediate important (sometimes inconvenient) truths about our fraught relationship with the worked land and remind us of the particular role of the creative arts in helping shape alternative, more sustainable relationships with the material resources available to us.

The tale of Flatford Mill

Our first miller's tale takes the form of a painting that visualises a particular, less predatory, aspect of the dynamic relationship between land (in the form of a crop) and water (in the form of a mill stream and imminent storm). Successive misreadings of John Constable's iconic Romantic canvas *The Hay Wain* have much to tell us about our inability to recognise the processes and priorities involved in milling as well as the embeddedness of the watermill within the contingencies of social and environmental history (see Figure 1.2). *The Hay Wain* depicts an unladen hay wagon in shallow water between Flatford Mill in East Anglia and the cottage of tenant farmer Willie Lott. The miller whose tale is told in this painting

Figure 1.2 John Constable, *The Hay Wain* (1821)

was the artist's landowning father, Golding Constable, who held the lease on Flatford Mill.

For many who will remember the image from place mats and biscuit-tin lids, *The Hay Wain* depicts rural life as a calm, bucolic idyll where the miller's seemingly contented labourers gather a picturesque crop of hay that has been drying on the fields. The scene is so pretty, the central character seems to have stopped work to admire his surroundings. Constable's painting has become synonymous with what we expect to see in historical portrayals of English village life. So much so that Jack Higgins's *The Eagle Has Landed* (1975), in which a waterwheel exposes a German plot to assassinate Winston Churchill, is set in a fictional Norfolk parish called Studley Constable. Karl Kroeber (1992) urges us to consider *The Hay Wain* as something other than a realistic representation of the processes and technologies involved in haymaking. Now more than ever, Kroeber warns, Constable's depiction of the Dedham Vale was, and is, 'addressed to an audience of non-haymakers' (1992: 29). The temptation is to view the canvas as an artful assembly of haymaking images, rather than as an accurate record of work. Kroeber quite rightly, in broad terms, suggests that *The Hay Wain* 'recall[s] our imagination to a function of our society that we have grown accustomed to overlooking' (1992: 29). Like most viewers, though, he also appears to have overlooked certain clues *vis-à-vis* farming techniques that were integral to haymaking in 1820, and which would certainly have been a matter of everyday experience for contemporary audiences.

Consider hay wains themselves. '[W]hat,' asks Kroeber, '*is* this one doing in the middle of the river?' (1992: 30). It's an interesting question, but is it the right question? In fact, the hay wain is not in the 'middle of the river,' but is positioned off the river Stour in a mill stream, which is something quite different. Leaving milling nomenclature aside, though, perhaps a more fruitful question to pose is this: where has the hay wain come from? Because if the wagon's destination seems clear enough – the fields being harvested by the stooped labourers in the right of the composition – the other end-point in its journey is less obvious, at least in the critical literature on the painting. The most often-mooted destination is Flatford watermill itself, which stands just behind the painting's viewing perspective. Ann Bermingham is not alone in asserting that the wagon has come from there: she writes of 'Flatford Mill from which the hay wain returns' (1989: 142). But in agricultural terms, this makes no sense. Grass and hay belong to pastoral agriculture, whereas the business of watermills is with arable produce such as wheat and barley. The most likely place from which the hay wagon is returning, as Roger Friedland and Deidre Boden (1994) point out, is a barn – there's one a little further behind the mill.[7]

The potential for such misreading arises only when we forget our historical relation with the worked land and, specifically, the relationship between its produce and water sources. Constable's portrayal of the latter has also posed problems among critics. Kroeber refers to the hay wain's crossing of the river Stour (and mill stream) as a 'short cut' (1992: 29). The route from the meadows and back again would have been the most direct and traditional route to the landowner's barn. Such journeys across fords are critical for the sustainability of communities, as reflected in the countrywide profusion of ancient place names in which the *ford* element occurs (Mills 2003). They also underlie a vexed legal ecosystem of leaseholds, tenancies, water rights and wages, as well as arrangements between tenant farmers and landowners, and in broader terms, local power structures, at the heart of which lay Constable's father.

While *The Hay Wain* might appear to transcend material history, offering an escapist's paradise, the work, as John Barrell points out, actually struggles to ignore, or absorb into its aesthetic, wide unrest and 'social divisions' affecting East Anglian agricultural communities (1980: 132). In 1816, just five years before the painting's composition, the region had witnessed 'bread or blood' food riots as a result of rising corn prices (Peacock 1965). Constable, in fact, commented dismissively on this unrest in an 1821 letter to his friend John Fisher – a letter in which he also discusses *The Hay Wain* (Leslie 1845: 90–3, 142–3, 145–6). It is perhaps reasonable to assume that the fact Constable's father owned the local granary might have inflected the painter's views on 'bread or blood'. The year 1822 saw a spate of hayrick and barn firings, resulting in transportations and executions for arson. Constable may be keen to paint, as it were, over the cracks in a community where his own family were influential landowners and merchants, but whatever calm the painting offers is only calm in the sense of that which precedes a storm – a literal storm.

According to the National Gallery's notes on the work, the hay wain appears to be serenely static, 'stand[ing] in the water' (National Gallery 2015). Agricultural labourers were rarely able to stand idle on the job, especially when there was a crop to be harvested. Water in a different form, the menacingly dark clouds in the upper left of Constable's painting, invites us to read the hay wain's short journey in a different context. Rain could spell disaster for a harvest of cut hay. What looks like the stationary pose of the central figure in fact shows him contemplating the imminent storm, and perhaps judging where it is headed, and how long it will take to arrive. He may also be considering whether the downpour is likely to make the mill stream impassable, and what he needs to do next in order to save as much of the hay harvest as possible. In the light of John Middleton's advice to farmers in *General View of the Agriculture of Middlesex*, the stasis many have seen in the central figure and hay wain is, in fact, a moment of tension and incipient action: 'In the very common case of approaching rain, when the hay is fit for carrying, every nerve is, or ought to be, exerted to … [get] all the carts and waggons loaded, and drawn into the barns' (1813: 316). The alternative was to leave the hay to 'take its chance' and to risk the valuable crop rotting and spoiling (Middleton 1813: 316). This very specific detail has wider ramifications. As Paul Muskett (1984) points out, the year after Constable finished his painting, heavy rain, which resulted in interrupted labour and lost wages, was a contributory factor in the East Anglian agrarian riots.

When we look closely, then, we see that Constable's painting is far from timeless and unchanging, but is forensically accurate in depicting the rhythms in motion of loading and unloading hay wagons. (There is a second wagon in the painting – often missed, tiny in the fields on the right, which is having stooks of hay lifted onto it. In visual 'dialogue' with the returning unladen wain, it again emphasises process rather than stasis.) The painting also insists on the importance of being able to 'read' water correctly, whether it takes the form of rainclouds or mill stream. Managing food (fodder for livestock) and water is shown to be essential to the livelihoods of the miller, his family (including the miller's gifted son) and employees. If we can read these material resources correctly, Constable's hay wain (and *The Hay Wain*) can be seen to exist at the centre of a fraught network of economic and social relations on the point of violent upheaval.

The tale of Trumpington Mill

Our second miller's tale, *The Reeve's Tale*, situates the working watermill in a particular geographical setting and historical moment. Chaucer's Reeve tells the story of Symkyn, the corrupt miller from the site of Tennyson's later inspiration, Trumpington near Cambridge. Like the thieving miller in the *General Prologue* of the *Canterbury Tales*, Symkyn has a 'thombe of gold' (Chaucer 2008: i.563). The miller's golden thumb is both metaphorical and literal, signifying the profit to be made by a talented miller who can winnow bad seed by hand, but also the use of cereal grains in determining the weight of gold and hence the currency, with one 'grain' being the weight of a grain of barley (later, a grain of wheat, which is lighter).[8]

Having defrauded his customers for many years, Symkyn gets his comeuppance, not by finally giving his customers what they are entitled to, but by suffering the indignity of having his wife and daughter fornicate with two students from the nearby university. The metaphor is plain and the economy of the story is the very definition of poetic justice: Symkyn grinds his customers' grain, and his customers grind his womenfolk. The satirical treatment of the miller, and the tale's concern with sex and social status, are responses to the power of the miller in medieval England. As guardians of the food chain, millers were regulated by the Assize of Bread (1266–1820) by which the government and local authorities regulated the ingredients, weight and price of bread (Davis 2004; Ross 1956). Chaucer's satire, and other aspects of this tale, suggest that in spite of the Assize, it was felt that the miller held a disproportionate amount of power.

In framing *The Reeve's Tale*, Chaucer's most arresting departure from his source text (a thirteenth-century French fabliau) is to locate Symkyn's mill in a very particular place: 'At Trumpyngtoun, nat fer fro Cantebrigge, / Ther gooth a brook, and over that a brigge, / Upon the whiche brook ther stant a melle' (Chaucer 2008: i.3921–3). The lines read like instructions to a traveller. Their specificity suggests that the tale the Reeve is about to tell is intimately related to this particular location. Chaucer's first readers are likely to have known this area, for in the immediate aftermath of the Peasants' Revolt (June 1381), the royal court was transferred to Cambridge. The relationship between the two places mentioned by the Reeve is one of antagonism, and for a very simple reason: the Cambridge colleges rely on the surrounding rural areas (including Trumpington) for their food. Within Cambridge itself, the river Cam was too weak to power a watermill, meaning the colleges had to take their grain elsewhere to be milled. Commandeering the spot at which the Cam flows with greatest strength, Symkyn has a monopoly over milling in the region, and this monopoly ('Greet sokene') is written into law:

> Greet sokene hath this millere, out of doute,
> With whete and malt of al the land aboute;
> And nameliche ther was a greet collegge
> Men clepen the Soler Halle at Cantebregge;
> Ther was hir whete and eek hir malt ygrounde. (Chaucer 2008: i.3987–91)

Universities are hungry places. Unable to either produce or process their own food, and within water rights, the Cambridge colleges were vulnerable to high prices and unscrupulous millers.

Chaucer's positioning of *The Reeve's Tale* in a particular landscape and historical moment has the potential to open up a new, politicised interpretation of the poem. Both before and after the Peasants' Revolt of June 1381, the rural areas of Cambridgeshire had been subject to violence and protest – largely by agricultural labourers angry at the erosion of their customary rights by government legislation and landowners (Aston 1994). During the feast of Corpus Christi in 1381, local labourers broke into Corpus Christi College and destroyed its property, burning many of its books. The symbolic weight of this act, taking place during a celebration of the 'host', the body of Christ prepared from the finest wheat by millers and bakers, cannot be underestimated. Thomas Walsingham (2003: 458–9) chronicled how, in nearby Hertfordshire, protesters refused to take their grain to St Alban's Abbey (also a site used by the royal court), which held a monopoly on milling rights in the region, and instead used hand mills. When the authorities seized the tenants' hand mills, protesters broke into the abbey, seized and smashed its millstone and distributed the fragments among rioters. By imitating the breaking of the host in communion, the protesters attacked the inequitable distribution of food and justice in the parish. Chaucer's *The Reeve's Tale* seems to echo and engage with these events, using Symkyn's watermill to dramatise the ongoing struggle between the competing interests of town and country, food producers and consumers.

The tale of Dorlcote Mill

The power afforded by water rights is key to Chaucer's study of food production, and our penultimate miller's tale shows what happens when millers, managers of river ways for many centuries, are threatened with the gradual erosion of their water rights. Telling the story of the Tullivers, a family of millers who have lived and worked at Dorlcote Mill on the fictional river Floss for several generations, most of the events described in George Eliot's novel, *The Mill on the Floss*, take place in the first two decades of the nineteenth century. Eliot is as careful as Constable and Chaucer in her depiction of a precise agricultural and environmental world. As at Flatford, Dorlcote Mill abuts arable land, which is also owned and worked by the Tullivers; their produce is distributed overland and by water. In the haunting opening chapter, Eliot's narrator remembers, as if in a vision or dream, the cornfields, orchards, mill and malt house owned by the Tullivers. This is, however briefly in the timeline of the novel, a sustainable model of a community which feeds itself.

The diversion of water upstream of Dorlcote anticipates and, it is implied, contributes to the eventual destruction of Tulliver's grain mill, home and, most importantly, the family's future, when the siblings Tom and Maggie Tulliver drown as freak storms coincide with the Floss's biannual tidal bore. Writing forty years after the events described, Eliot's narrator remembers the mill just as its role as centre of the sustainable community is vanishing. The ensuing analysis of the reasons for this loss are sharply critical and insightful:

> Our instructed vagrancy, which has hardly time to linger by the hedgerows, but runs away early to the tropics, and is at home with palms and banyans – which is nourished on books of travel and stretches the theatre of its imagination to the Zambesi – can hardly get a dim notion of what an old-fashioned man like Tulliver felt for this spot, where all his memories centred, and where life seemed like a familiar smooth-handled tool that the fingers clutch with loving ease. (Eliot 2010: 299)

Why has the sharp satire of Chaucer's miller turned to elegy? Because of 'Our instructed vagrancy'. Eliot explains that we have lost a vital connection with the land because we no longer know how to occupy and take responsibility for it. Serving as a metaphor for the rootless individual, the food chain has also become ungovernable. In 1820, the Assize of Bread had just passed out of legislation. The authorities no longer undertook to guarantee the price, ingredients and size of bread. Millers, having lost their powerful role in the community, were pitiable subjects of tragedy rather than satire.

Eliot's analysis of sustainable food production, and the pressures it faced, goes much further. As we have seen, the 1820s were times of food riots and rural violence – customary rights were not to be given up without a fight. This violence, which hovers at the edges of *The Mill on the Floss*, is most obviously manifested in the flood and drowning of Maggie and Tom Tulliver at the end of the novel. Critics have tended to neglect the very careful research Eliot undertook into such matters as water management, arable farming and the laws pertaining to property and water rights. Informed by this knowledge, the tragic flood is far from a 'melodramatic contrivance' (to quote E. A. Baker (1937: 247)), but is shown to be the consequence of abandoning the mill as the centrepiece of sustainable food production. This trend in criticism of the novel is summarised and discussed by Larry Rubin (1956). Elsewhere, we have argued that Eliot's knowledge of the agri-environmental world inhabited by the Tullivers was acquired first-hand (Archer et al. 2015a). In particular, she was interested in different types of river systems, including mill streams, tidal rivers and waterways with contested water rights, and her research included trips to the river Wey (and at least one of its mills) and Gainsborough, Lincolnshire, on the river Trent. In the novel, Eliot carefully documents the way in which the flood is the result of the failure in river management. The destructiveness of the flood is shown to be a consequence of the emergence of mills (paper, cotton, oil and iron mills) upstream, the conversion of arable land into meadow and pasture, and because grain mills such as Dorlcote are purchased by the likes of Pivart, a 'new name'. The Tullivers owned Dorlcote Mill 'a hundred year and better' (Eliot 2010: 174); new names, it is evident, do not have the knowledge, passed down the generations, to manage their land and water. The result is catastrophic:

> Nature repairs her ravages – repairs them with her sunshine, and with human labour. The desolation wrought by that flood had left little visible trace on the face of the earth, five years after. The fifth autumn was rich in golden corn-stacks, rising in thick clusters among the distant hedgerows; the wharves and warehouses on the Floss were busy again, with echoes of eager voices, with hopeful lading and unlading. (Eliot 2010: 598)

What is taken by critics as a 'hopeful' ending means something very different to those more attuned to the fragility of the food chain. 'The fifth autumn' sees an eventual restoration of cereal crops. But four autumn and five spring harvests have failed – impossible conditions for those who used to live on the Floss and work its land. Trading towns like St Ogg's can endure a short period of dearth, but for the innumerable smallholding families like the Tullivers – and the workforce which relied on them – five years of dearth signals the destruction of the self-sustaining community.

Conclusion: food at the watershed

It is one of the themes of this essay that, throughout history, the coming together of flowing water and harvested food, in the shape of the grain mill, has been decisive for sustaining thriving communities in time and space. In recent years, mill restoration projects have become widespread and even fashionable, driven in part by the heritage industry and by interest in hydropower as an alternative energy source. The Community Heritage Fund (financed by the Landfill Tax) instigated the Windmill and Watermill Challenge in 2007 and awarded £600,000 to eight watermill and windmill projects across England. More recently the Heritage Lottery Fund has made contributions to a few watermill conservation schemes, including Howsham mill in Yorkshire and the Sacrewell mill near Peterborough (Society for Protection of Ancient Buildings 2009). Many other restorations have been community or private initiatives.

One such venture is Felin Ganol (Figure 1.3), a centuries-old watermill site in the Ceredigion village of Llanrhystud, which until the Second

Figure 1.3 Felin Ganol, Llanrhystud, Ceredigion

World War was producing flour and ground animal feeds, and generating electricity through an Armfield turbine. The mill's website describes the history of Felin Ganol and its current produce (Felin Ganol Watermill 2015; see also O'Sullivan 2014). The mill was central to village life as a source not only of cereals and electrical power but also of apples from its orchard, of milk and butter from its cows, of livestock including pigs, hens and ducks reared for sale by the miller's family, of tea from an importing business operating from mill premises, and of timber from a carpenter in the mill loft who used a saw driven by the turbine.

Relative isolation meant that the revolution in agricultural practices that occurred in the eighteenth and nineteenth centuries in England did not have much impact on the rural way of life in Wales until well into the last century. Topography and climate make cereal production precarious. Almost 60 per cent of land in Wales is above 500 feet and unsuitable for arable cultivation. Even in the areas where cereals were grown, the absence of large grain markets inhibited widespread wheat cultivation in favour of livestock and oats (Moore-Colyer 2000). Measures such as liming to improve soils relied on deliveries by canal and sea (Moore-Colyer 1988). Sustaining a food chain under these circumstances is difficult at the best of times. As the twentieth century progressed, major changes in lines of communication and in popular tastes and aspirations contributed to the opening up of formerly self-sufficient communities and the loss of the mill's social and economic status. Felin Ganol fell idle as the demand for stoneground flour dwindled.

Recommissioning derelict grain mills should also be considered in the context of the growth of the local food movement. In this respect, Felin Ganol has a significant story to tell. Anne and Andy Parry bought Felin Ganol in 2006 and set about rescuing the neglected water supply, wheel gear, grindstone mechanisms and the grain handling and flour grading machinery. The fully functioning mill now produces and sells wholemeal, white, semolina, spelt, rye and triticale flours and is organically certified. Since September 2010 Felin Ganol has been milling the spring wheat variety Tybalt, grown locally on Aberystwyth University Organic Farm, and is part of a project with Aberystwyth and Bangor Universities to develop naked oat and barley varieties specifically suited for growing in the mid-Wales area. The Parrys represented Wales at the 2014 Salone del Gusto and Terra Madre in Turin and presented Hen Gymro wheat to the Slow Food Ark of Taste, a project created in 1996 to defend endangered foods from globalisation. At a time when the risks to food security and sustainability associated with grossly extended and vulnerable supply lines make headline news, the re-establishment of long-defunct local food chains is welcome.

The watermill has been a site of intense scrutiny in British culture. The stories we have told about these structures form part of the larger narrative of our own lost connection with the land and the literature embedded in it. They continue to be sustaining, meaningful presences in our lives. The vigorous literary stream, flowing from William Langland through Chaucer, Shakespeare, John Clare, George Eliot and Thomas Hardy, has petered out in the swamps of *Cold Comfort Farm* and *Scoop*; a chronicle, like the history of mills, milling and millers, of the failing sustainability of a tradition. By recovering the lost contexts of the mill as the heart of self-sufficient communities throughout history, we stand a better chance of understanding how we got here, and how we may prepare, as sustainably as we are able, for a future that seems to be brewing up a perfect storm of food, energy, biodiversity and climate calamity.

Notes

1 William Blake writes of mills in 'There is No Natural Religion' (1788), 'The Marriage of Heaven and Hell' (1790–3), 'America a Prophecy' (1793), 'Milton a Poem' (1804–11), 'Jerusalem the Emanation of the Giant Albion' (1804–20) and his letter to William Hayley, 23 October 1804 (Eaves et al. 1996).
2 Britain is also figured as Ceres in the illustrated frontispiece to *Poly-Olbion* (Michael Drayton 1612, 1622).
3 The agrarian crisis of the late sixteenth and early seventeenth centuries is documented by Andrew McRae (2002), R. B. Outhwaite (1991) and John Walter (1991). Literary and cultural representations of dearth and food riots in this period are discussed in Elisabeth Angel-Perez and Alexandra Poulain (2008) and Chris Fitter (2000).
4 Gordon Williams lists literary examples of this usage between 1525 and 1720 (Williams 1994: ii.887–9).
5 See, for example, Burgundy's speech at the end of *Henry V* (1598–9: 5.2.33–62), discussed in Archer et al. (2013: 532–3).
6 'The Miller's Maid: A Tale' was first published in Bloomfield's 1801 publication *Rural Tales* (Bloomfield 1857: 121–9). It was made into an opera by John Davy in 1804 and formed the basis for a two-act melodrama by John Faucit Saville in 1821.
7 For the literary and technological significance of barns, see Marggraf Turley (2014).
8 This is discussed by Bee Wilson (2008: 64) in her history of adulteration and swindling in the food supply chain.

References

Angel-Perez, Elisabeth and Alexandra Poulain (eds) 2008. *Hunger on the Stage.* Cambridge: Cambridge Scholars Publishing.

Archer, Jayne Elisabeth, Richard Marggraf Turley and Howard Thomas 2013. 'The Autumn King: Remembering the Land in *King Lear*', *Shakespeare Quarterly* 63 (4): 518–43.

Archer, Jayne Elisabeth, Richard Marggraf Turley and Howard Thomas 2014. *Food and the Literary Imagination*. Basingstoke: Palgrave Macmillan.

Archer, Jayne Elisabeth, Richard Marggraf Turley and Howard Thomas 2015a. '"Moving Accidents by Flood and Field": The Arable and Tidal Worlds of George Eliot's *The Mill on the Floss*', *English Literary History* 82 (2): 701–28.

Archer, Jayne Elisabeth, Howard Thomas and Richard Marggraf Turley 2015b. 'Reading Shakespeare with the Grain: Sustainability and the Hunger Business', *Green Letters: Studies in Ecocriticism* 19 (1): 8–20.

Aston, Margaret 1994. 'Corpus Christi and Corpus Regni: Heresy and the Peasants Revolt', *Past & Present* 143: 3–47.

Baker, Ernest Albert 1937. *The History of the English Novel, Vol. 8: From the Brontës to Meredith*. n. pl.: Barnes & Noble.

Barnes, Donald Grove 1930. *A History of English Corn Laws: From 1660–1846*. London: Routledge.

Barrell, John 1980. *The Dark Side of the Landscape: The Rural Poor in English Painting, 1730–1840*. Cambridge: Cambridge University Press.

Bermingham, Ann 1989. *Landscape and Ideology: The English Rustic Tradition, 1740–1860*. Berkeley and Los Angeles, CA: University of California Press.

Bloomfield, Robert 1857. *The Poetical Works of Robert Bloomfield*. Ed. Birkett Foster. London: Richard Clay.

Butler, Joseph 1726. *Fifteen Sermons Preached at the Rolls Chapel*. London: Printed by W. Botham for James and John Knapton.

Byron, George Gordon Lord 1823. *The Age of Bronze*. London: John Hunt.

Camden, William 1610. *Britannia, or A Chorographicall Description of the Most Flourishing Kingdomes, England, Scotland, and Ireland, and the Ilands Adioyning*. Trans. Philemon Holland. London: Georgii Bishop & Ioannis Norton.

Capecchi, Danilo 2013. 'Over and Undershot Waterwheels in the 18th Century. Science-Technology Controversy', *Advances in Historical Studies* 2: 131–9.

Cervantes, Miguel de 2005. *Don Quixote*. Trans. Edith Grossman. London: Vintage.

Chambers, E. K. 1923. *The Elizabethan Stage*. 4 vols. Oxford: Clarendon Press.

Chaucer, Geoffrey 2008. *The Riverside Chaucer*. Ed. Barry D. Benson. Oxford: Oxford University Press. 3rd edn.

Davis, James 2004. 'Baking for the Common Good: A Reassessment of the Assize of Bread in Medieval England', *Economic History Review* 57 (3): 465–502.

Delasanta, Rodney 2002. 'The Mill in Chaucer's *Reeve's Tale*', *Chaucer Review* 36 (3): 270–6.

Eaves, Morris, Robert N. Essick and Joseph Viscomi (eds) 1996. The William Blake Archive. www.blakearchive.org/. Accessed 13 February 2017.

Eliot, George 2010. *The Mill on the Floss*. London: Vintage.

Felin Ganol Watermill 2015. www.felinganol.co.uk/. Accessed 13 February 2017.

Fitter, Chris. 2000. '"The Quarrel is Between our Masters and us their Men": *Romeo and Juliet*, Dearth, and the London Riots', *English Literary Renaissance* 30 (2): 154–83.

Fletcher, John and William Rowley 1909. *The Maid in the Mill*. In Francis Beaumont and John Fletcher, *The Works of Francis Beaumont and John Fletcher*, vol. 3. Ed. A. R. Waller. 10 vols. Cambridge: Cambridge University Press.

Friedland, Roger and Deidre Boden 1994. 'NowHere: An Introduction to Space, Time and Modernity'. In *NowHere: Space, Time and Modernity*. Ed. Roger

Friedland and Deidre Boden. Berkeley and Los Angeles, CA; London: University of California Press, 1–60.

Gash, N. 1978. 'After Waterloo: British Society and the Legacy of the Napoleonic Wars', *Transactions of the Royal Historical Society* 28: 145–57.

Hobbes, Thomas 1985. *Leviathan*. Ed. C. B. Macpherson. London: Penguin Books.

Jefferies, Richard 1880. 'The Water-Mill. Field Names'. In Richard Jefferies, *Round about a Great Estate*. London: Smith, Elder and Co., 163–82.

Kaplan, Steven S. 1984. *Provisioning Paris: Merchants and Millers in the Grain and Flour Trade during the Eighteenth Century*. Ithaca, NY: Cornell University Press.

Keats, John 1988. *The Complete Poems*. Ed. John Barnard. Harmondsworth: Penguin Books. 3rd edn.

Kroeber, Karl 1992. *Retelling/Rereading: The Fate of Storytelling in Modern Times*. New Brunswick, NJ: Rutgers University Press.

Leslie, C. R. 1845. *Memoirs of the Life of John Constable*. London: Phaidon. 2nd edn.

Levin, Harry 1959. *A Gallery of Mirrors: Observations on Novelists and Poets*. Cambridge: Riverside Publishing.

McRae, Andrew 2002. *God Speed the Plough: The Representation of Agrarian England, 1550–1660*. Cambridge: Cambridge University Press.

Marggraf Turley, Richard 2014. 'Agritech and the Arts: From Barns to D-Space', online lecture delivered at the 31st Chaos Communication Congress (#31C3), Hannover, 29 December. http://tinyurl.com/pjrzg3a. Accessed 13 February 2017.

Marggraf Turley, Richard, Jayne Elisabeth Archer and Howard Thomas 2012. 'Keats, "To Autumn", and the New Men of Winchester', *Review of English Studies* 64: 797–817.

Mekonnen, M. M. and A. Y. Hoekstra 2010. 'The Green, Blue and Grey Water Footprint of Crops and Derived Crop Products', *Hydrology and Earth System Sciences* 15: 1577–600.

Middleton, John 1813. *General View of the Agriculture of Middlesex*. London: Sherwood, Neely, and Jones. 2nd edn.

Mills, A. D. 2003. *A Dictionary of British Place-Names*. Oxford: Oxford University Press.

Moore-Colyer, Richard 1988. 'Of Lime and Men: Aspects of the Coastal Trade in Lime in South-West Wales in the Eighteenth and Nineteenth Centuries', *Welsh History Review* 14: 54–77.

Moore-Colyer, Richard 2000. 'Wales'. In *The Agrarian History of England and Wales VII (Part 1) 1850–1914*. Ed. E. J. T. Collins. Cambridge: Cambridge University Press, 427–52.

Muskett, Paul 1984. 'The East Anglian Agrarian Riots of 1822', *Agricultural History Review* 32: 1–13.

National Gallery 2015. 'The Hay Wain'. www.nationalgallery.org.uk/paintings/john-constable-the-hay-wain. Accessed 13 February 2017.

O'Sullivan, Eve 2014. 'Meet the Flour Producer: "A Typical Day is Much the Same as it Would Have Been 150 Years Ago"', *Guardian*, 4 January 2014. www.theguardian.com/lifeandstyle/2014/jan/04/meet-the-flour-producer-felin-ganol. Accessed 13 February 2017.

Outhwaite, R. B. 1991. *Dearth, Public Policy and Social Disturbance in England, 1550–1800.* Cambridge: Cambridge University Press.

Peacock, James Alfred 1965. *Bread or Blood: A Study of the Agrarian Riots in East Anglia in 1816.* London: Gollancz.

Pinion, F. B. 1984. *A Tennyson Companion.* London: Palgrave Macmillan.

Reynolds, Terry S. 1983. *Stronger than a Hundred Men: A History of the Vertical Water Wheel.* Baltimore, MD: Johns Hopkins University Press.

Rodner, W. 1997. *J. M. W. Turner: Romantic Painter of the Industrial Revolution.* Los Angeles, CA: University of California Press.

Ross, Alan S. C. 1956. 'The Assize of Bread', *Economic History Review* 9 (2): 332–42.

Rowland, Beryl 1969. 'The Mill in Popular Metaphor from Chaucer to the Present Day', *Southern Folklore Quarterly* 33 (33): 69–79.

Rowland, Beryl 1970. 'The Oven in Popular Metaphor from Hosea to the Present Day', *American Speech* 45 (3/4): 215–22.

Rubin, Larry 1956. 'River Imagery as a Means of Foreshadowing in *The Mill on the Floss*', *Modern Language Notes* 71: 18–22.

Society for the Protection of Ancient Buildings 2009. 'The Mills Section', www.spab.org.uk/spab-mills-section/. Accessed 13 February 2017.

Stiglitz, Joseph E. et al. 2009. 'Report by the Commission on the Measurement of Economic Performance and Social Progress'. http://library.bsl.org.au/jspui/bitstream/1/1267/1/Measurement_of_economic_performance_and_social_progress.pdf. Accessed 23 March 2017.

Swade, Doron 2002. *The Difference Engine: Charles Babbage and the Quest to Build the First Computer.* London: Penguin Books.

Tucker Brooke, C. F. (ed.) 1908. *A Pleasant Comedie of Faire Em, The Millers Daughter of Manchester: With the Loue of William the Conquerour.* In *The Shakespeare Apocrypha.* Ed. C. F. Tucker Brooke. Oxford: Clarendon Press, 118–56.

Walsingham, Thomas 2003. *The St Albans Chronicle: The Chronica Majora of Thomas Walsingham, 1376–1394,* Volume 1. Ed. and trans. John Taylor, Wendy R. Childs and Leslie Watkiss. Oxford: Oxford University Press.

Walter, John 1991. 'The Social Economy of Dearth in Early Modern England'. In *Famine Disease and the Social Order in Early Modern Society.* Ed. John Walter and Richard Schofield. Cambridge: Cambridge University Press, 75–128.

Watts, Martin 2006. *Watermills.* Princes Risborough: Shire Publishing.

Williams, Gordon 1994. *A Dictionary of Sexual Language and Imagery in Shakespearean and Stuart Literature,* 3 vols. London: Athlone Press.

Wilson, Bee 2008. *Swindled: The Dark History of Food Fraud from Poisoned Candy to Counterfeit Coffee.* Princeton, NJ: Princeton University Press.

2

Sustenance from the past: precedents to sustainability in nineteenth-century literature and culture

John Parham

Introduction: sustainability has no history

In her 1980 book *The Death of Nature: Women, Ecology and the Scientific Revolution*, Carolyn Merchant argues both that 'new social concerns generate new intellectual and historical problems' and that 'new interpretations of the past provide perspectives on the present and hence the power to change it' (1980: xvi). While this offers a rationale for the study of 'Victorian ecology', the question of whether sustainability even has a history is vexed.

There is little indication, for example, of historical precedents in the book most often credited with enshrining the principles of sustainability in the environmental movement – E. F. Schumacher's *Small is Beautiful* (1973). There's little or nothing in academic studies of sustainability or, correspondingly, of sustainability in major cultural histories of environmentalism or ecology such as Donald Worster's *Nature's Economy* (1994). Indeed, some of those histories (though not Worster's) are ambivalent about any nineteenth-century tradition of green ideas. David Pepper's survey, *Modern Environmentalism* (1996), identifies Romanticism and Victorian ecological socialism (e.g. the work of William Morris) as precursors of contemporary environmentalism but neglects sustainability and is guarded about Romanticism in particular. Likewise, Andrew Dobson, in the fourth edition of *Green Political Thought* argues strongly against any correspondence between the long Romanticist tradition and a contemporary ecological thinking whose specific elements, he argues, simply were not there in the nineteenth century, whether anxiety over 'environmental crisis' or the organic, systemic paradigms that underlie scientific and philosophical ecology (2007: 22–6). And yet Derek Wall's *Green History* (1994) has a chapter on 'Sustainable Development' that includes three nineteenth-century writers – Percy Shelley, George Perkins Marsh and the French utopian socialist François Fourier; John Stuart Mill, in *Principles of Political Economy* (1920 [1848]), wrote about 'the stationary state';

while Morris envisaged something akin to a sustainable society in *News from Nowhere* (1993 [1890]).

Writing in the *Observer* newspaper, the political journalist Andrew Rawnsley (2014) recently compared current divisions amongst the British Conservative Party over Europe to what he regards as the party's last 'fundamental' schism when, in 1846, it was critically divided by the Corn Laws. He writes, 'History does not repeat itself, but it can rhyme'. Resisting (for the most part) a temptation to play with the relationship between the Corn Laws and conceptions of sustainability, I will nevertheless develop my argument via Rawnsley's notion of rhyming. Namely, that while ecological sustainability was not literally there as a concept in the nineteenth century, contemporary patterns of thought did anticipate it.

The introduction to this volume makes clear the impossibility of regarding sustainability as a unified discourse. While arguing that a distinct understanding of sustainability can be seen emerging from nineteenth-century cultural and literary history, nevertheless I will also suggest that this evidences and illuminates the complexities, contingencies and competing priorities the term contains. Sustainability's history was constituted by the pan-European emergence of a set of conditions, responsive philosophical paradigms and literary texts. In a sense this has already been historicised by Ulrich Grober's 2010 book *Sustainability: A Cultural History*. While markedly (though not entirely) Germanic, Grober includes a chapter on the eighteenth and nineteenth centuries, entitled 'The Birth of Ecology', from which a three-part framework for the cultural development of sustainability can be gleaned. Working from Grober's account, my opening section will consider where and in what form sustainability emerged in the nineteenth century. Retaining a focus on that century, the middle section will explore, more generally, how literature can articulate principles of sustainability. Lastly, the essay closes with an analysis of two novels – *News from Nowhere* (1890) and Emile Zola's *La Terre* (1887). In their different ways, these books can inform our understanding of how sustainability might be conceptualised, narrated, even practised.

Sustainability's history

The first of three elements emphasised in Grober's history is that sustainability was founded upon a philosophical paradigm that, contrary to the 'death of nature', emphasised the energy, complexity and autonomy (from humans) of nature. He refers, essentially, to the Romantic paradigm of Vitalism, which stresses that phenomena in nature have 'independent powers' of self-generation, animation, self-direction and the ability to act (Packham 2012: 1). From there Grober develops a trajectory that encompasses, as it does for environmental historians like Wall and Worster, Linnaeus' *The Oeconomy of Nature* (1749). Here the classification of plants and the sense of a (God-given) balance of species is driven, Grober argues, by 'sheer delight' in multitude, diversity, fertility (2012: 90). Grober's trajectory extends to Johann Gottfried Herder's *Outlines of a Philosophy of the History of Mankind* (1800), which formulates the Earth and its biosphere as bound to the physical forces of the atmosphere and sun (Grober 2012: 96). And from there it reaches towards the physicist Alexander von Humboldt, whose works, such as *Cosmos* (1845), more firmly established the web of connections between flora, fauna, environment and geography as 'an integral whole given animation and direction by inner forces' (cited in Grober 2012: 104).

One can append a parallel English context that Eric Wilson characterises as 'romantic turbulence': Wordsworth's 'sense sublime' of a world 'deeply interfused' in 'Tintern Abbey'; or Coleridge's view, developed in *Hints Toward the Formation of a More Comprehensive Theory of Life* (1816; published 1848), that life discloses itself through a 'principle of individuation' which, nevertheless, reveals 'unity in the many' (see Wilson 2000: 23, 100). Likewise, as in Germany, one might trace this trajectory onto Victorian literature and culture: Carlyle's receptivity, even anticipation, of energy physics in *Sartor Resartus* (see Myers 1989); Ruskin's realisation that 'life' is characterised by a radical interconnectedness that encompasses perpetual nascence and decay (see Bardini 2014: 10); or Hopkins's poetic embodiment, via the deployment of a stress-based 'sprung rhythm', of ecological relations structured by energy exchange (Parham 2010: 53).

Yet, for Wilson, 'Goethe is the exemplar' of this epistemology (2000: xviii); for Grober too he is the 'fulcrum' of the 'convoluted story' (2012: 108) of the origins of sustainability. Influenced by Linnaeus, and encapsulating the Weimar Classicism that surrounded himself and Herder, Goethe crystallised this emergence of an understanding of nature as possessed of ceaseless vitality, autopoiesis, power, and beauty. He is also fundamental to the second aspect of Grober's analysis, namely that the idea of sustainability derives, specifically, from a consequent deliberation as to how human activity – land use, industry, development, social structures – answers to a nature unceasingly dynamic and turbulent. Heather I. Sullivan has written that Goethe's philosophy and writings precede, or anticipate, a contemporary 'open-systems' model which posits that organisms co-exist in a state of dialectical 'affinity' whereby each being is simultaneously made and remade by an environment (2011: 244–5) on which it, itself, nevertheless acts. Consequently, for Sullivan *Faust* articulates the blindness of human 'consciousness' when it 'sees its own agency but not its inevitable affinities' (2011: 246) but also the realisation that if we recognise our dual existence and affinity with other nature, then we can, ourselves, effect alterations that serve a mutually-supportive, sustainable environment (see Sullivan 2011: 248).

Central, of course, to Romanticism's own emergence was a reaction to science, industry and, more generally, modernity. This leads to the third point Grober makes in documenting the history of sustainability: in the context of the origins of ecology, sustainability developed both as a discourse which articulated an anxiety about whether we, humans, could sustain ourselves while the project of civilisation or modernity continued on its way and as an attempt to find practical solutions by which we might sustain human being. One of the particularly interesting aspects of Grober's analysis is the revelation that, invariably, the philosophical speculations described above were underpinned by practical considerations. The cultural context underlying Linnaeus' *The Oeconomy of Nature* was the defeat of Sweden in the Great Northern War in 1721. Rampant Swedish militarism in central Europe removed the peasants from the land and left the 'corn ... rotting in the fields'; defeat consigned the Swedes back to their own territory and 'restricted to their own resources'. So, 'In order to adapt to the new situation they imported the doctrine of cameralism from the German princely states complete with its ideological structure of self-sufficiency and self-government, careful use of one's own resources, and sustainability' (Grober 2012: 92). When Sweden experienced famine in 1756 Linnaeus offered up a list of edible native flora as alternative food sources.

Likewise, active in civic duties, in later life Goethe wrote that 'it was only the desire to be able to offer ... practical advice', on the use of the woods, parks and gardens around the ducal palace at Weimar which 'drove me to study nature'. Goethe's preoccupation, as Grober puts it, was primarily with how a 'tiny, resource-poor, famine stricken ... territory in the middle of Germany' might develop 'on the basis of its own resources' (2012: 94). Correspondingly, it is noted that early in his career Humboldt worked for the Prussian mining authorities in the mountains in Bavaria, that one of his first tasks was to address a wood shortage, and that later he was angered that deforestation in the tropical rainforests of Venezuela had led to shortages of wood and water (Grober 2012: 104). The Vitalist conception of nature was founded, in other words, in the context of a wider examination of human capabilities, human labour, and the kinds of society one should work towards (Grober 2012: 107). Sustainability developed, if you like, as the social or political wing of human ecology.

Questioning why the word 'ecology', coined by Ernst Haeckel in 1866, remained in the cultural background until the late twentieth century, Grober suggests that the practical connotations raised by the Romantic conception of a living, powerful nature were buried under a competing, increasingly dominant free-market liberalism. For example, the briefly fashionable doctrine of cameralism had advocated State-administered strategies for achieving sustainable self-sufficiency in the supply of food and raw materials via measures like environmental improvement, reclamation or the indigenous cultivation of hitherto imported crops (see Jonsson 2013: 55–6). Yet liberalism dismissed such early philosophies of sustainability as the 'outdated doctrine[s] of petty central European states' (Grober 2012: 103; and see 106). While this might explain why sustainability (and its rhyming philosophies) remains invisible even in most environmental histories, as Goethe wrote, 'Where there is much light, there will also be strong shadow' (1799 [1773]: 17). The purpose of this historical analysis is, then, to examine the role that literature can and did play, within these counter-histories, both in drawing a philosophy of sustainability out of the shadows and in modelling and articulating its implications for human society and culture. As will now be considered, Grober's analysis suggests that both the concept of sustainability and its literary articulation emanated from an emerging idea of 'environment'.

Sustainability and the concept of environment

Goethe and Herder studied nature together in the 1780s. Grober suggests that, for each of them, related concepts of 'evolution', 'development' and 'environment' became increasingly important (2012: 107). 'Environment', in particular, was the concept which engendered an understanding that human life is dictated by our surroundings even as we, in turn, act upon them. Occurring across Europe, and across disciplinary boundaries – science, philosophy, social science, culture – this growing consciousness of 'environment' was facilitated by literature. That fact – and ultimately literature's capacity to represent sustainability – can be traced via the emergence and definition of interlocking terms: *environment* (Britain), *milieu* (France), *Umwelt* (Germany), and *ambiente* (Italy). Returning to *milieu* later, I will, for now, focus upon the interrelationship of the English and German terms.

In English, *environment* was first used as a noun by Carlyle in his translation of a passage in Goethe's autobiography where Goethe discusses the influence of surroundings on his personality. This, writes Grober, 'is, so to speak, the birth certificate of a word' (2012: 110). *Umwelt* emanates from Goethe's *Italian Journey* (1816–17), though Goethe appears to have picked it up from the Danish poet and travel writer Jens Baggesen (Spitzer 1942: 207; Grober 2012: 108). Foregrounding the ecological importance of the concept, Grober touches upon the inflection of *Umwelt* as it was developed by the Estonian German naturalist Jakob von Uexküll in his 1909 book *Environment and Inner World of Animals*. Recently, von Uexküll has had a great influence on the fields of biosemiotics and ecocriticism.

His definition best captures the significance of the developing concept of environment to that of sustainability.

Umwelt was promulgated by von Uexküll to express a subjective sense of environment. It refers, on the one hand, to the totality of everything the organism perceives and internalises from its environment and, on the other, via the supporting concept of *Innenwelt*, to the influence of the mind and, subsequently, human practice on that environment (see Grober 2012: 110; Wheeler 2011: 124). Wendy Wheeler writes:

> All organisms, von Uexküll argued, live in *Umwelten* which are signifying environments characterised by semiotic loops flowing ceaselessly between the *Umwelten* (semiotic environments) and *Innenwelten* (semiotic 'inner worlds') of creatures: each making each in a ceaseless living ecological process ... What a creature (as instance of a species) recognizes, or knows (and compares), are the signs in its environment which are necessary to its survival (and, thus, to its species' survival). And, of course, this applies to humans too. (Wheeler 2011: 272)

The concept of *Umwelt*/environment captures, then, a dynamic model, one that prefigures the contemporary ecological paradigm of *emergence.* Emergence posits that organisms and species evolve within an environment that shapes them but that that environment is, in turn, reshaped as each constituent organism evolves. Species and environments emerge and (re) emerge in tandem. In terms of a cultural understanding of sustainability, this means two things: as organisms partially shaped by our environment, we are (or ought to be) compelled both ecologically and existentially towards assessing how far the particular nature of human activities, social organisation and prevailing ideologies, discourses and cultural constructions (not least, sustainability) will, in their actual or likely environmental impacts, contribute towards sustaining human being; second, the fundamental role played by *Innenwelt* highlights that a significant part of that assessment will be carried out culturally.

Leo Spitzer's classic 1942 essay, 'Milieu and Ambiance: An Essay in Historical Semantics', sheds light on where and how the latter occurred. Focusing predominantly on the French variant of these terms, Spitzer traces the development of the phrase *milieu ambient. Milieu ambient* is generally defined, Spitzer writes, as 'the element immediately surrounding a given body' (1942: 173). He demonstrates how this concept evolved across biology, social science and poetics. The first, the biological definition, immediately signals the relevance of this discussion to our purposes: the 'surrounding element', he writes, 'is that which environs, not an inert substance, as in physics, but a living being; and so *milieu ambient* represents the element in which an organism *lives* and upon which it depends for sustenance' (1942: 175). Spitzer then describes how 'an ever-growing feeling of the solidarity existing between man and nature' (1942: 175–6) resulted in a convergence of the natural and social sciences. Consequently, *milieu ambient* gained a sociological dimension which extended the biological definition to incorporate the sum of 'exterior circumstances' shaping human lives, including society, politics and economics.

A fundamental issue posed, however, by the all-encompassing aspect of (natural *and* social) 'exterior circumstances' was the seemingly deterministic subjugation of all life to a rigid *milieu*. Such determinism undermines, as we shall see, a crucial component of sustainability. A qualification arose, Spitzer suggests, with a further revision of *milieu ambient* that began when the largely spatial sense of 'environing' was supplemented, in a development he attributes to the French physiologist Claude Bernard, with an interior dimension (Spitzer 1942: 182) not unlike the *Innenwelt*. That itself could be deterministic, as suggested in more pessimistic accounts of ideology such as Marcuse's 'One Dimensional Man'. Yet here, Spitzer asserts, what was inculcated was a sense (implicitly, and semantically, ecological) that humans belong to their *milieu* 'not as a captive to his jailor, but as a man to his home' (1942: 186).

Spitzer outlines how this steady, semantic conceptualisation of 'environment' in the nineteenth century gradually engendered an awareness that, while environmental conditions were partly attendant on how we ourselves modified the environment, this in turn was shaped by the ways in which, philosophically or culturally, humans modelled environments. The crucial development occurred when the word *ambient* evolved from being, in effect, a somewhat static noun (*milieu ambient*) to encapsulate more dynamically the qualities that animate an environment. This happened when, semantically, the word took on a new form – *ambiance* (see Spitzer 1942: 199). Embedding a sense of environments as mutable, *ambiance* foregrounded that our surroundings are (in part) constituted by how we imagine them. While this would gradually help foster an anticipation of sustainability as a fundamental prerequisite for human existence, more immediately it liberated the emergent paradigm of 'environment'. Spitzer writes: 'ambiance entered upon a brilliant career … in the literary language, a word evocative of a spiritual climate or atmosphere, emanating from, hovering over, a milieu' (1942: 188). In the beautifully apt 'hovering', he extols *ambiance* as 'the antithesis of the deterministic milieu' (Spitzer 1942: 191). This would lead, in places, to a 'conception of Western man as triumphant (at least in comparison with his brother of the East) over his environment' (Spitzer 1942: 186). That in turn would have ramifications in Nazism or, more recently, in free-market liberalism. Yet this remained qualified because, Spitzer states, *ambiance* was in actuality never entirely free of *milieu* (1942: 199). Rather, what the newly created semantic eminence of *ambiance* constituted was precisely the equipoise between environmental determinism and human agency that Grober sees as integral to the dialectic of sustainability that emerged in the nineteenth century. And such dialectical complexity rendered the concept of 'environment' ripe for literary treatment.

In damaging environmental systems, we endanger the very thing that determines human existence. Conversely, by reconceptualising environments, through creative representations that alter our sense of a given environment's ambience, we have the agency to mitigate those dangers. Literature, in the nineteenth century, was seen to have this capacity for altering conceptions and offering possibilities because its modes – whether non-discursive (e.g. poetic) or discursive (e.g. narrative) – could carry this complex interweaving of *milieu* and *ambience*, determination and agency. Working in the intellectual currents described above, John Stuart Mill, for example, suggested that the representation of such complex interrelations requires the ability, which he attributed to Carlyle, both to intuit 'many things … not visible' and to offer, in turn, a 'poetry to animate' (Mill 1920: 132–3). The relevance of this to contemporary discourses of sustainability – heavily implied, given that Mill posited the concept of 'The Stationary State' – is indicated by the fact that this nineteenth-century interplay between *milieu*/environment and ambience has echoes in recent debates in ecocriticism.

Environment, literature, sustainability

In his landmark ecocritical study, Lawrence Buell (1995) established a much-cited, four-point definition of what constitutes an 'environmental text'. Most relevant here are his first and fourth points: that 'The nonhuman environment is present not merely as a framing device but as a presence that begins to suggest that human history is implicated in natural history'; and that environmental texts are characterised by 'some sense of the environment as a process' (Buell 1995: 7–8). In response to the accusation, however, that Buell's criteria run the risk of coming too close to literary mimesis (see Phillips 2003: 6–9), one might summon Timothy Morton's 'ecomimesis', founded, Morton writes, on an 'ambient poetics' (2007).

For Morton, we are enmeshed in nature as ecological beings and can only hope to make sense of that condition through social or cultural constructions. Accordingly, we should seek not mimetic depictions of environments but, rather, ones that represent 'a circumambient, or surrounding, *world* … something material and physical' but nevertheless 'intangible' (Morton 2007: 33; author's italics). Such an aesthetic, were it to be developed with reference to Spitzer, might encapsulate what Morton regards as the 'strangeness' of one's environment (2007: 34). Seeing the 'openness', fluidity and lack of certainty of non-representational art (from Coleridge to The Cure) as well-placed to articulate this 'strange' ecomimetic ambience, Morton argues that we need a 'dark ecological' aesthetics (rather than idealisations of nature) to confront an estrangement from nature that has always been part of the human condition but which has now been heightened by environmental crisis. Morton suggests that the distancing achieved, by shattering our illusions, would be a necessary step towards politicising ecological sentiment. It would compel us to attempt to work out how, in practical terms, we might co-exist with nature (see Morton 2007: 113).

Commentators such as Kate Soper (2011) have accused Morton of evading what those forms of relationship might be and of lacking any tangible solutions or strategic political vision. Consideration of how an eco-mimetic ambient aesthetics might be developed in more practical (social or political) directions, such as sustainability, turns us towards the novel, a connection Spitzer made in relating the interplay of *milieu* and ambience to Balzac, the Goncourt brothers and Zola. In his 1998 article 'Problems in Ecocriticism and the Novel', Dominic Head suggests that 'If biological science can definitively break down the separation between the human individual and his or her environment ... then the implications for our modes of perception appear to be enormous' (1998: 69). In that context, he argues, an ecological aesthetics would require 'The apparent paradox of a representation which is stylized yet referential' (Head 1998: 67). Such an aesthetic is available within the textuality of the novel. For structural and narrative discontinuities, the ambiguities of diction or focalisation are well suited, Head suggests, to articulating the discordant, dialectical continuity between the human and her or his environment (1998: 67, 69). As Serpil Oppermann puts it, the novel has a unique potential to 'disclose how the discursive constructions of nature shape and condition the human valuation and understanding of the environment' (2008: 243).

In this light, ecocritics have made claims for contemporary, deconstructive forms such as the modernist or postmodern novel (see Heise 2002; Oppermann 2008; Rozelle 2002). As Head and others have pointed out, the same arguments might be applied, however, to the intricate textuality of the classic realist and naturalist novels of the nineteenth century (Head 1998: 66; Parham 2011: 25–7). Martin Ryle, for instance, has suggested that the temporal basis of these novels might allow for a sustained analysis of the emergence, evolution and impact of both natural and ecologically destructive processes (2000: 12). Such novels could hold the capacity to map both environmental diminution and/or the emergence of future, eco-utopian and (perhaps) sustainable societies. With these points in mind, I will, in the remainder of this essay, consider *News from Nowhere* and *La Terre*, two virtually contemporaneous novels, embedded in the same trans-European developments that Grober and Spitzer describe. The two novels indicate not only that discursive constructions akin to sustainability had a presence in nineteenth-century literature, but also how these constructions could now inform discourses of sustainability.

Morris and the sustainable novel

In many ways the most evident nineteenth-century novel about sustainability, *News from Nowhere* emerged out of the shadow history that Grober describes. As Peter Gould argues, as a prominent member of the Social Democratic Federation, Morris represented an 'early green politics' which sought to modify a prevailing Marxist emphasis on industrial production with the values of decentralised, land-based economies (see 1988: 31, 62–3). While the origins for this lay in Morris's British Romanticism (notably the influence of Keats – see Thompson 1977), such a perspective also appears to have had roots in what Anna Vaninskaya has called 'Germania.' This extolled Northern European models of rural, democratic, communal and subsistence living, an influence that Morris apparently shared with Marx and Engels, and seemingly derived from the social and intellectual context (not least the doctrine of cameralism) described by Grober (see Vaninskaya 2010: 185–7).

News from Nowhere both critiques Victorian society and imagines and formulates a more ecologically viable alternative. Yet the force of Morris's depiction of the social and ecological demise of Victorian industrial society is not met by an equivalently rich description of a future utopian society. His utopia seems sterile and is imagined through clichéd reference points: clean, clear water, and salmon swimming in an unpolluted Thames; urban London rebuilt on Italian architectural lines; the countryside as a garden cultivated by all; entertainment afforded by Welsh folk songs. Morris's utopia lacks, that is, any animating ambience. This is particularly evident in comparison to both the pungent metaphor by which he describes Victorian England – 'the beetle gets used to living in dung; and these people, whether they found the dung sweet or not, certainly lived in it' (Morris 1993: 125) – and the vivid narrative as 'Old Hammond' recounts a Victorian insurrection. Reminiscent of an adventure novel, Morris's novel describes how 'a glittering officer on horseback came prancing out from the ranks' and announced an order to disperse that was ignored by the crowd:

> I saw three little machines being wheeled out in front of the ranks, which I knew for mechanical guns. I cried out, 'Throw yourselves down! They are going to fire!' But no one scarcely could throw himself down, so tight as the crowd were packed. I heard a sharp order given, and wondered where I should be the next minute; and then – It was as if the earth had opened, and hell had come up bodily amidst us. It was no use trying to describe the scene that followed. Deep lanes were mowed amidst the thick crowd; the dead and dying covered the ground, and the shrieks and wails and cries of horror filled all the air, till it seemed as if there was nothing else in the world but murder and death ... How I got out of the Square I scarcely know: I went, not feeling the ground under me, what with rage and terror and despair. (Morris 1993: 144)

The relative failure of the utopian vision is not because of any limitations as a writer! Rather, *News from Nowhere*, initially serialised in 1890 in the Socialist League's journal *Commonweal* (of which Morris was editor), arguably should not be regarded as a novel at all (Pinkney 2010: 99). It was a political narrative – a blueprint for utopia – with as much a pragmatic as literary purpose. While it probably does highlight why we need a 'poetry to animate' our visions of sustainability, its limitations also reveal something fundamental about sustainability itself. Grober complains that, preoccupied with pollution and environmental degradation, the word 'environment' has lost its complexity and vitality, as described by Spitzer. As a consequence, we invariably downgrade nature 'to a threatened "setting" or "surrounding" for human life' (2012: 110–11), precisely what Buell suggests an 'environmental text' should not do. Conversely, what *News from Nowhere* demonstrates, in evidencing the difficulty of an ambient eco-aesthetics of sustainability, is that the ideal sustainable society does not, and cannot, exist; it is, indeed, 'nowhere'.

Morris knew this. Highlighting a contrast between 'Old Hammond' who, though functioning as the conscience and keeper of the utopian society's 'customs, values and memories', is, in fact, 'disappointed' with life (Morris 1993: 99) and the (anti-entropic) vitality brought about by the appearance, late on in the novel, of twenty-year-old Ellen, Tony Pinkney reads these narrative developments as suggestive of the fact that revolutions, socio-ecological or otherwise, are 'never definitively won' (2010: 101). The utopian, sustainable society has to be continuously renegotiated, reconstituted and evolved, a precise analogy to the concepts of dialectical systems of affinity and perpetual environmental emergence that constitute scientific ecology. This leaves us – and Pinkney credibly argues that Morris perceived this – with a more radical and realistic conception of sustainability as a permanent, ongoing project, working with a nature that we cannot ever definitively manage. A utopian sustainable society, crudely understood, is then just as much a management of nature as those lamented by environmental critics of modernity. This was also Zola's understanding. His representation of an existing society articulates both the anxiety that arises from the dialectical struggle to live within nature, an anxiety strained further by modernity, and the fundamental terms and conditions on which human life can be sustained.

Novels of sustainability

Zola was expressly preoccupied with issues akin to contemporary discourses on sustainability. One could cite the portent in his mining novel *Germinal*, 'Hasten to be just, or the earth will open up beneath our feet' (quoted in Nelson 2007: 14). Conversely, his later work was marked, Julia Przybos has argued, by social utopianism, notably a belief that Malthusian predictions of exponential population growth could be solved by scientific discovery and technological innovation (2007: 169; 178–9). Przybos finds this in *Fécondité* (1899), where Mathieu Froment practises irrigation and utilises technology to bring fallow land and marshland into productivity (Przybos 2007: 181). Having suggested in *Fécondité* that there need be no poverty, Zola's next novel, *Travail* (1901), addresses the forms of social organisation that might bring this about. Influenced by Fourier, Zola depicts the creation of *phalanstères*, small-scale socialist communities of roughly 5,000 acres cultivated by about 1,600 people. While Przybos suggests that such an idea 'still resonates within the rising culture of alternative solutions' (2007: 185), Brian Nelson has described *Travail* as 'highly didactic' (2007: 17). Certainly, the novel, like *News from Nowhere*, is at times schematic and lacking in vitality. Having addressed, through Morris, the utopian novel, I want here then to turn my attention to *La Terre*, part of the Rougon-Macquart cycle. For this novel connects the idea of human sustenance as an ongoing project more closely to a recognisable, contemporary everyday life.

Arch-representative of a literary naturalism defined by its preoccupation with historical, social and environmental determinism (White 2011: 524), Zola's interest in these ideas was influenced by three figures, all cited by Spitzer – Balzac, Claude Bernard and the critic and historian Hippolyte Taine. Taine exerted a particular influence on Zola's literary ideas via his paradigm of *race*, *milieu* and *moment* (national and family heredity; social and biological environment; historical point in time). The third of Taine's categories, *moment*, is of particular interest. Its emphasis on temporality potentially introduces historical change into the paradigm, and, therefore, a suggestion that attributes such as subjectivity, agency and intervention might temper social or environmental determinism. Noting, however, that Taine had little feel for nature beyond a mechanistic sense of 'forces conditioning human life' (Spitzer 1942: 178), Spitzer nevertheless concludes that the paradigm still tended towards the overly deterministic:

> *Le moment* is not superfluous in Taine's system, but represents a recognition of the necessity to take into account the imponderable. It was, however, only a partial recognition, and neither the introduction of this term, nor his salutary 'hesitations' were adequate to off-set his still too rigid adherence to naturalistic parallels. (Spitzer 1942: 178)

This did not apply to Zola. The preface to the second edition of *Thérèse Raquin* famously documents Zola's preoccupation with social and environmental determinism. Yet Douglas Parmée, in an introduction to *La Terre*, has argued that these social scientific enthusiasms had waned by the time Zola wrote this particular novel (in Zola 1980: 5). Likewise, Susan Harrow suggests that while Taine's model did influence the Rougon-Macquart cycle, *moment* became an increasingly prominent component

(2007: 106–7). Accordingly, for all the social philosophy etched into the structure of *La Terre*, the resultant combination of environmental determination and human agency, alongside the animating force of Zola's desire to create 'the living poem of the Earth, but in human terms, not symbolically' (Parmée, in Zola 1980: 11), engendered the perfect *terrain* on which to explore the three key elements found in Grober's outline of the historical development of sustainability. For in *La Terre*, Zola acknowledges and explores nature and the environment as an agential living force; translates this into a corresponding examination of how humans could exist and sustain themselves within ceaselessly shifting, emergent environments; and encapsulates anxieties contained within discourses around sustainability.

La Terre

An understanding of the Earth as a living force is animated via a metaphor that compares the land, the Plain of Beauce, where the novel is set, with the sea, the archetype of living, autonomous nature:

> At first there was nothing to see on the broad brown fields but barely perceptible touches of green along the ground. Then this tender green grew bolder, more velvety, and became almost uniform in colour. Then the wisps of corn grew and thickened out until each plant took on its special hue; he could pick out from afar the yellowy green of wheat, the blue-green of oats, the grey-green of the rye, in fields stretching out in all directions as far as the horizon, amid the red patches of clover. This is the time when Beauce is lovely, dressed in youthful spring attire, uniform and refreshing to the eye in its monotony: the stalks grow longer and turned into a sea, a sea of grain, heaving and deep and limitless ... a gentle wind would blow in steady gusts, hollowing the fields out into waves which started on the skyline and swept along until they died away on the further horizon. The fields quivered and grew paler, the wheat was shot through with tints of old gold, the oats were tinged with blue whilst the rye trembled with glints of purple. And as one undulation followed the next the heaves heaved ceaselessly under the ocean breath. (Zola 1980: 200–1)

The ambient, oceanic metaphor vitalises the flat landscape. Correspondingly, focalised through the farmer Buteau's newly enchanted eye (having just acquired the land), the colours, though 'barely perceptible', make this life force visible to both mind and soul. Nevertheless, having expressly desired to render this 'living poem of the Earth ... in human terms', Zola is clear that the plain of Beauce, otherwise a 'bare plateau', emerges in this vitality and beauty only through the integral human activity of farming. Hence the vision at the very end of the novel of 'the infinite expanse of the rich plain of Beauce swarming with sowers, swinging their arms in the same monotonous gesture' (Zola 1980: 500). For Zola, human sustenance is dependent upon productive action within a given environment that we are 'forced to cultivate ... in order not to starve' (1980; 499). Humans and environment (e)merge together: affinitive, co-existent.

Even prior to the long description above, Zola had, though, introduced a cluster of anxieties as to the continuation of the sustenance provided by the earth. Framed by a concern that the land around Beauce was becoming 'exhausted' and 'infertile' (1980: 153), three particular anxieties emerge. These concern: the capaciousness of nature; the capaciousness of human nature; and the specific impact of modernity. Immediately following the passage above, Buteau recalls the prevalence of storms and the possibility, reversing his analogy, that a 'raging sea' might leave his crops 'razed to the ground' (Zola 1980: 201). (Conversely, it is a drought that, soon after, almost wipes out his crops). Correspondingly, while Zola continues to invoke the mutuality of humans and earth, an earlier chapter, which has further wave-like images of reapers 'lunging forward without a pause, all in the same rhythm, their bodies swaying from the hips and their scythes swinging steadily to and fro' (Zola 1980: 141), nevertheless highlights the regularity with which reaping and haymaking are interrupted. For throughout the novel, indolence, lack of interest, greed, jealousy, alcohol, or lust interfere with the necessary work of cultivating the earth.

Addressing the extent to which this perceived exhaustion of the land has been exacerbated by human indolence, the novel contains ongoing discussions about how to increase yield. Generally articulated by the progressive farmer Hourdequin, possible solutions include chemical fertilisers, methods of crop rotation, soil analysis, accounting and mechanisation (see Zola 1980: 153–4, 156). A further discussion (echoing Mill) occurs around the respective merits of small-scale and industrial farming. This remains unresolved, the potential benefits of the labourer having (like Buteau, at this stage) a personal investment in the land offset by the increased hard labour, decreased economies of scale and impossibility of deploying machinery on small plots. Subsequently, while Hourdequin's potential remedies represent much the same faith in modernity that Zola endorses in *Fécondité*, the impression here is of a perplexed, despairing search for solutions. What that implies is modernity's essential impotence in terms of conserving sustainability, partly as a result of human imperfection but also because of the problems that modernity itself has created. Equivalent to what Rawnsley invokes, Corn Law type battles between free trade and protectionism (Zola 1980: 152–3) raise the spectre of globalisation extinguishing local farming and commerce. And other modern factors compound the fears haunting this novel. Throughout, people are drifting from the land and from an understanding of how it sustains us – whether from lack of interest, the vivid temptations of the city, or the relative ease (it would appear) of making money by trading in foreign commodities. *La Terre* demonstrates an anxiety about the land continuing to sustain its human inhabitants that resonates through contemporary discourses of sustainability.

Conclusion

News from Nowhere and *La Terre* both suggest a literature that can nourish, envision and enrich sustainability, not didactically, but in the complex 'environmental' sense that informed the writing of Grober, Goethe, Mill or Taine. The question of how literary texts might nourish contemporary articulations, even practices, of ecologically sustainable living is currently under discussion. Most notably, Hubert Zapf has pressed the claims for literature exploring issues such as sustainability when he asserts that, while 'culturally embedded', literary writing also offers 'distinct form[s] of textuality and knowledge' engendered by its unique 'codes and autopoetic rules of production' (2016: ch. 2). For Zapf, literature's distinctiveness and complexity – forged, for example, from narrative, imagery, characterisation – offers an appropriate fit with the 'complex processes of interaction in the living world of material nature'. It is of value, specifically, to sustainability because of an ability to invoke 'longer-term perspectives of evolution and survival beyond short-term economic interests', as in *La Terre* (see Zapf 2016: ch. 3). However, while positing literary texts as 'a living force field of transformative energies', potentially able to revitalise our understanding of ideas such as sustainability, Zapf also insists that realising that transformative potential will 'depend on the active participation' of readers bringing their own meanings to the text. Interestingly, this idea of transformation arising ultimately from the 'creative activity' of readers leads him to speculate that the 'creative energies of texts can travel across periods and cultures' (Zapf 2016: ch. 2).

The question of whether the novels discussed above realise Zapf's contention that literary texts might open up a recognition of the environmental consequences of our actions and/or new perspectives and possibilities would require further analysis, possibly even reader-reception studies. Nonetheless, the presence of both Morris and Zola in recent ecologically oriented discussions of utopianism (see Pinkney 2010; Przybos 2007) – and of Morris, and Zola's inspiration (Fourier), in Wall's *Green History* – suggests that these writers can be or are already regarded as important precursors. Furthermore, the analysis above indicates that these books' distinctively literary modes do help untangle the radically dialectical sense of human–environment relations that emerged across Europe in the nineteenth century. The novels afford an 'aesthetic transformation' of the cultural discourses that Grober charts and offer, in Zapf's terms, a 'metadiscursive space' (2016: ch. 2) out of which a contemporary reader might discern important lessons about sustainability.

The lessons might be summarised as follows: that the discourse of sustainability which has arisen since the 1960s is not a historically specific response to a particular crisis (e.g. anthropogenic climate change or the growing excesses of consumer society) but one component within an ongoing response to the pressures exerted by modernity; that there exists a long cultural tradition which understands that ecological imperatives form an enduring basis to human life; that those imperatives render sustainability a political issue and raise questions about the structure of society; that, however, a sustainable society will 'never definitively [be] won' and constitutes, instead, something to aspire to; lastly, that the project of sustainability is ongoing precisely because it involves working with the forces of a 'nature' that we will never definitively manage.

In this final sense *La Terre*, in particular, delivers a deeper, more profound understanding of the complexities that surround sustainability but also the indisputable reality that governs human being – we need to sustain ourselves within 'nature'. Writing in this volume, Hannes Bergthaller suggests that the paradox of a sustainability which insists on limits and constraints while simultaneously seeking to transform human lifestyles and behaviours need not be seen as a weakness. On the contrary, this dialectic of two competing priorities undergoing constant renegotiation is precisely why sustainability constitutes 'genuinely political matter'. Goethe and Herder's dual sense of nature's perpetual transformation and permanence is strikingly similar. I hope, therefore, to have demonstrated both that sustainability can be historicised and that something approaching contemporary ecological notions of sustainability emerged as identifiable (if unnamed) concepts from the intellectual, social and literary currents of nineteenth-century (post)Romanticism.

The nineteenth-century history of sustainability mapped in this chapter also corroborates that sustainability, as a concept, never stands still. It is, as several essays in this book testify, endlessly discursive. Even when colonised by ideology, 'sustainability' remains a flexible, agile paradigm, perpetually political, and so perpetually historical. And yet; though offering an ongoing responsiveness and vigilance to the conditions by which, at any given moment, we live within 'nature', sustainability also issues an ongoing reminder that we live under constraints dictated by the nonhuman. In that sense, a more useful concept might, arguably, be *sustenance*: making choices (e.g. between technological and organic food production) and creating conditions for sustaining ourselves and the nonhuman nature with which we co-exist and on which we rely. Zola's remedy, for example, to his society's anxiety about sustenance, was to understand the importance of working with the land, even though aided by the judicious (sustainable) application of new technologies and methods. This, he believed, would offset Malthusian concerns about 'exhaustion' and about the conservation of humanity.

Referring to the fire which destroyed what had been Hourdequin's farm, Zola writes: 'The walls might be burnt down but you couldn't burn down the earth. Mother earth would always be there to feed those who sowed her. She had space enough and time, until people learned how to make her produce more' (1980: 499). Yet in keeping with the counterbalance, in sustainability, of constraint, the final key point to arise from the novel is that an autonomous, self-organising, ever-emergent earth (and 'Earth') will, in the end, continue on its own path. Regardless of human intervention, we are reminded, there is every chance that nature, as an independent force, will carry on regardless:

> It was said that the land would pass into other people's hands and harvest from other countries would overwhelm ours and all our fields would be overgrown with brambles. So what? Can you harm the earth .... The earth doesn't take part in our petty, spiteful, antlike squabbles, she pays no more attention to us than to any other insects, she merely goes on working and working, eternally. (Zola 1980: 499)

In *La Terre* Zola returns us to a world in which sustainability, or sustenance, is an ongoing, dialectical, and contingent project. In this light, when we talk about sustainability, we are also talking about human sustainability, a point we'd do well to remember.

References

Bardini, Thierry 2014. 'Decompicultures: Decomposition of Culture and Cultures of Decomposition', *Green Letters: Studies in Ecocriticism* 18 (1): 9–21.

Buell, Lawrence 1995. *The Environmental Imagination: Thoreau, Nature Writing, and the Formation of American Culture*. Cambridge, MA and London: Belknap Press.

Dobson, Andrew 2007. *Green Political Thought*. London: Routledge. 4th edn.

Goethe, Johann Wolfgang von. 1799 [1773]. *Gortz of Berlingen with the Iron Hand. An Historical Drama of the Fifteenth Century*. Trans. Rose Lawrence. Liverpool: n. pub.

Gould, Peter 1988. *Early Green Politics: Back to Nature, Back to the Land and Socialism in Britain 1880–1900*. Brighton: Harvester.

Grober, Ulrich 2012 [2010]. *Sustainability: A Cultural History*. Trans. Ray Cunningham. Totnes: Green Books.

Harrow, Susan 2007. '*Thérèse Raquin*: Animal Passion and the Brutality of Reading'. In *The Cambridge Companion to Zola*. Ed. Brian Nelson. Cambridge: Cambridge University Press, 105–20.

Head, Dominic 1998. 'Problems in Ecocriticism and the Novel', *Key Words* 1: 60–73.

Heise, Ursula 2002. 'Toxins, Drugs, and Global Systems: Risk and Narrative in the Contemporary Novel', *American Literature* 74 (4): 747–78.

Jonsson, Fredrik Albritton 2013. *Enlightenment's Frontier: The Scottish Highlands and the Origins of Environmentalism.* New Haven & London: Yale University Press.

Merchant, Carolyn 1980. *The Death of Nature: Women, Ecology and the Scientific Revolution.* London: Wildwood House.

Mill, John Stuart 1920 [1848]. *Principles of Political Economy with Some of their Applications to Social Philosophy.* Ed. W. J. Ashley. London: Longmans, Green, & Co.

Morris, William 1993 [1890]. *News from Nowhere and Other Writings.* Ed. Clive Wilmer (Penguin Classics). London: Penguin.

Morton, Timothy 2007. *Ecology without Nature: Rethinking Environmental Aesthetics.* Cambridge, MA and London: Harvard University Press.

Myers, Greg 1989. 'Nineteenth-Century Popularizations of Thermodynamics and the Rhetoric of Social Prophecy'. In *Energy and Entropy: Science and Culture in Victorian Britain.* Ed. Patrick Brantlinger. Bloomington and Indianapolis, IN: Indiana University Press, 307–38.

Nelson, Brian 2007. 'Zola and the Nineteenth Century'. In *The Cambridge Companion to Zola.* Ed. Brian Nelson. Cambridge: Cambridge University Press, 1–18.

Oppermann, Serpil 2008. 'Seeking Environmental Awareness in Postmodern Fictions', *Critique* 49 (3): 243–53.

Packham, Catherine 2012. *Eighteenth-Century Vitalism: Bodies, Culture, Politics.* Basingstoke: Palgrave Macmillan.

Parham, John 2010. *Green Man Hopkins: Poetry and the Victorian Ecological Imagination.* Amsterdam and New York: Rodopi.

Parham, John 2011. '"For You, Pollution": The Victorian Novel and a Human Ecology. Disraeli's *Sibyl* and Gaskell's *Mary Barton*', *Green Letters* 14: 23–38.

Pepper, David 1996. *Modern Environmentalism: An Introduction.* London: Routledge.

Phillips, Dana 2003. *The Truth of Ecology: Nature, Culture, and Literature in America.* Oxford and New York: Oxford University Press.

Pinkney, Tony 2010. 'Version of Ecotopia in *News from Nowhere*'. In *William Morris in the Twenty-First Century.* Ed. Phillippa Bennett and Rosie Miles. Oxford: Peter Lang, 93–106.

Przybos, Julia 2007. 'Zola's Utopias'. In *The Cambridge Companion to Zola.* Ed. Brian Nelson. Cambridge: Cambridge University Press, 169–87.

Rawnsley, Andrew 2014. 'The Battle for What it Means to be on the Right in British Politics', *Observer* 31 August.

Rozelle, Lee 2002. 'Ecocritical City: Modernist Reactions to Urban Environments in *Miss Lonelyhearts* and *Paterson*', *Twentieth-Century Literature* 48 (1): 100–15.

Ryle, Martin 2000. 'Place, Time and Eco-Criticism', *Green Letters* 2: 11–12.

Schumacher, E. F. 1993 [1973]. *Small is Beautiful: A Study of Economics as if People Mattered.* London: Vintage.

Soper, Kate 2011. Review of 'Tim Morton, *The Ecological Thought*', *Radical Philosophy* 165: 55–7.

Spitzer, Leo 1942. 'Milieu and Ambiance: An Essay in Historical Semantics', *Philosophy and Phenomenological Research* 3 (2): 169–218.

Sullivan, Heather I. 2011. 'Affinity Studies and Open Systems: A Nonequilibrium, Ecocritical Reading of Goethe's *Faust*.' In *Ecocritical Theory: New European Approaches*. Ed. Axel Goodbody and Kate Rigby. Charlottesville, VA: University of Virginia Press, 243–55.

Thompson, E. P. 1977. *William Morris: Romantic to Revolutionary*. London: Merlin Press.

Vaninskaya, Anna 2010. 'William Morris's Germania: The Roots of Socialism.' In *William Morris in the Twenty-First Century*. Ed. Phillippa Bennett and Rosie Miles. Oxford: Peter Lang, 169–92.

Wall, Derek (ed.) 1994. *Green History*. London: Routledge.

Wheeler, Wendy 2011. 'The Biosemiotic Turn: Abduction, or, the Nature of Creative Reason in Nature and Culture.' In *Ecocritical Theory: New European Approaches*. Ed. Axel Goodbody and Kate Rigby. Charlottesville, VA: University of Virginia Press, 270–82.

White, Nicholas 2011. 'Naturalism.' In *The Cambridge History of French Literature*. Ed. William Burgwinkle, Nicholas Hammond and Emma Wilson. Cambridge: Cambridge University Press, 522–30.

Wilson, Eric 2000. *Romantic Turbulence: Chaos, Ecology and American Space*. New York: St Martin's Press.

Worster, Donald 1994. *Nature's Economy: A History of Ecological Ideas*. Cambridge and New York: Cambridge University Press. 2nd edn.

Zapf, Hubert 2016. *Literature as Cultural Ecology: Sustainable Texts*. London: Bloomsbury Academic Open Access. www.bloomsburycollections.com/book/literature-as-cultural-ecology-sustainable-texts/ch2-the-ecocultural-potential-of-literature. Accessed 13 February 2017.

Zola, Emile 1980 [1887]. *La Terre [The Earth]*. Trans. Douglas Parmée. London: Penguin.

3

Deep sustainability: ecopoetics, enjoyment and ecstatic hospitality

Kate Rigby

Mind the gap!

In 2007 an article appeared in the science journal *Trends in Ecology and Evolution* with the witty title, 'Mind the Sustainability Gap'. The gap in question refers especially to the ecological dimension of the sustainability agenda and concerns the chasm that continues to yawn ever wider between 'what we know needs to be done and what is actually being done' to avert catastrophic climatic and environmental change (Fischer et al. 2007: 621). While the authors acknowledge 'regional-scale improvements in some indicators of poverty, food supplies and the environment', they argue that these are 'overshadowed by ongoing deterioration of key biophysical indicators at the global scale', especially with regard to biodiversity loss and global warming (Fischer et al. 2007: 621). In view of the evident failure of existing approaches to sustainability to redress such dire threats to more-than-human life on Earth, Fischer et al. recommend a redirection of sustainability research, policy and management along two main axes. Firstly, in place of the conventional 'triple bottom line' of environmental, social and economic sustainability, they favour a 'hierarchical' model, with the 'biophysical limits of Earth setting the ultimate boundaries within which social and economic goals must be achieved' (Fischer et al. 2007: 621) Secondly, they argue that closing the sustainability gap necessitates bridging the disciplinary divide:

> Human action in the world emerges from a complex dialectic among the living world itself, the social contexts of human life and action, and the conceptualisations through which human life is made meaningful. Fundamentally enhanced collaboration among natural and social scientists and scholars of human contexts, symbols and meanings would signal the beginning of a new paradigm for addressing the sustainability gap. (Fischer et al. 2007: 623)

In this chapter, I argue that the twofold renovation of the concept of sustainability proposed by Fischer et al. invites a deeper questioning of prevailing cultural assumptions, perceptions and values regarding human identity, aspirations and interrelations with nonhuman others and our earthly environs, and I explore the potential contribution of literature to this cultural work of 'deep sustainability'.

The 'transdisciplinary' research programme advocated by Fischer et al. is modelled by the article itself, which arose from a multi-perspectival workshop on sustainability hosted by the Australian National University's Fenner School of Environment and Society. Its seventeen co-authors include physicists and ecologists, geographers and engineers, agricultural scientists and conservation biologists, along with the co-founders of Australia's National Working Group in the Ecological Humanities, which had been inaugurated at the Australian National University in 2001: historian of science and environment, Libby Robin, and anthropologist and cultural theorist, Deborah Bird Rose. Among the other workshop participants thanked in the acknowledgements was the eminent feminist ecophilosopher, Val Plumwood, and it is from her paper, 'Deep Sustainability as Cultural Work', that this chapter takes its title. Following Plumwood's death in 2008, an article partially based on that paper was published in the Ecological Humanities Corner of the *Australian Humanities Review*, under the revised title 'Nature in the Active Voice'. Here, Plumwood differentiated her depth model of sustainability from conventional constructions of both 'deep ecology', with its prioritisation of 'wilderness' preservation, and 'shallow ecology', with its privileging of exclusively human interests. Instead, she proposed a 'mixed framework' that reveals how 'human-centredness can have severe costs for humans as well as non-humans' (2009: 116). Rejecting the 'pernicious false-choice' of the deep/shallow divide, Plumwood argues that human-centredness – 'a complex syndrome which includes the hyperseparation of humans as a special species and the reduction of non-humans to their usefulness to humans, or instrumentalism' – engenders a hazardous 'failure to understand our embeddedness in and dependency on nature [and] distorts our perceptions and enframings in ways that make us insensitive to limits, dependencies and interconnections of a non-human kind' (2009: 116). Accordingly, in her earlier paper, she had argued that the 'cultural work of deep sustainability' entailed the critical investigation of conceptual frameworks and social systems that occlude the agency and interests of nonhuman others, along with the 'ecological services', upon which human social and economic sustainability remain dependent. In 'Nature in the Active Voice', she goes on to suggest how certain forms of writing can help to loosen modern cultures out of the bonds of human self-enclosure by providing a space for what she calls an 'animating sensibility and vocabulary' (Plumwood 2009: 126) that recognises other-than-human creative agencies, communicative capacities and ethical considerability. This she had previously hailed as a 'critical green writing project' that 'might make visible whole new interspecies dialogues, dramas and projects', and thereby 'dispel the sado-dispassionate "imaginary" … that has supported and nourished the post-enlightenment illusion of human monopoly of mindful, cultural, intentional elements in the world' (Plumwood 2007: 19).[1]

In her workshop paper, Plumwood explicated 'sustaining' in line with earlier ecological feminist articulations of an ethics of flourishing (e.g. Cuomo 1998), namely as referring to activities that 'nourish' or 'support', thereby 'contributing to the other's resilience and flourishing' (Plumwood 2006: 1). Following this definition, an apt emblem for sustainability might be found in the figure of the nest: an avian work of ecopoetics, understood literally as the making of an *oikos*, a place and a practice for bringing new life into being, which, in its interwoven threads of diverse materials, reiterates the connective processes that compose flourishing ecosystems. In the latter part of this chapter, I will return to the nest, specifically as it figures in the work of the Romantic poet John Clare and is refigured in the ecopoetic experimentation of the contemporary writer (and erstwhile conservation biologist) David Morley.

The interpretive frame that I bring to this discussion is informed by several further lines of theorisation which enrich Plumwood's proposal for a 'radical green writing project': ecophilosopher Freya Mathews's transpecies ethic of 'bioproportionality' (2014), which I relate to Derrida and Dufourmantelle's notion of 'radical hospitality' (2000), and Mathews's 'ontopoetic' model of the transvaluation of desire (2010), which I connect with recent work on 'alternative hedonism' by Kate Soper and others (Soper et al. 2009). While the trajectories of radical hospitality and alternative hedonism are seemingly divergent, with the former presupposing an altruistic ethic of alterity as opposed to the self-pleasuring implicit in the latter, I argue that both are necessary to advancing the work of deep sustainability and can be shown to bear upon one another in the ecopoetic practices that I discuss here. Firstly, though, it is necessary to consider more closely the limitations of conventional understandings of sustainability, in order to explain why, and how, these shortcomings need to be redressed.

Reframing sustainability: beyond the 'triple bottom line'

As previously indicated, one of the two major problems with prevailing models of sustainability research, policy and management identified by Fischer et al. arises from the construction of sustainability as consisting of three 'pillars'. This metaphor features influentially in the UN General Assembly's resolution endorsing the outcome of the 2005 summit on sustainable development, which included the commitment to promoting 'the integration of the three components of sustainable development – economic development, social development and environmental protection – as interdependent and mutually reinforcing pillars' (2005: 11–12). While 'pillars' invokes an architectural image in which the removal of any one support structure would cause the ceiling to collapse, another way of imaging sustainability along these lines is the Venn diagram, in which the three dimensions are seen to be distinct, but overlapping at a central point. This has given rise to the popular concept of the 'triple bottom line', which implies, firstly, that these are discrete concerns, and secondly, that there is a parity between them. In much corporate and governmental practice, this has enabled economic considerations to take precedence, often primarily in the interests of a privileged minority, moreover, with matters of social development and environmental protection being addressed with more or less token measures, if at all. In many cases, then, the semantically slippery rhetoric of sustainability is deployed with a view to maintaining capitalist business-as-usual in 'developed' nations, while extending it to 'developing' ones. It is for this reason, then, that Fischer et al. call for the reconceptualisation of the three components as a hierarchy of considerations, based on the recognition that '[s]ocieties cannot exist without a functioning life-support system, and economics can only flourish within a functioning social system with effective institutions and governance structures' (2007: 622). In other words, there is only one bottom line, and it is set by those biophysical processes that have engendered, and remain crucial to, the diverse more-than-human life of this planet.

This recommendation echoes some earlier formulations of sustainability (e.g. Eichler 1999) that have since been sidelined, but are now being rearticulated in relation to the UN's proposed new Sustainable Development Goals. In a 2013 Comment in *Science*, for example, the eminent climate scientist, Dave Griggs, in company with a group of other researchers (including Will Steffen, one of the co-authors of Fischer et al. 2007), proposed a revised model of sustainability based on what they termed a 'nested concept' (Griggs et al. 2013: 306). This entails an amendment to the widely accepted definition of the UN's 1987 World Commission on Environment and Development (chaired by Gro Harlem Brundtland), which is verbally modest, but conceptually momentous. Instead of framing sustainable development as 'development that meets the needs of the present without compromising the ability of future generations to meet their own needs', the proposed new definition refers to 'development that meets the needs of the present while safe-guarding Earth's life-support system, on which the welfare of current and future generations depends' (World Commission 1987: 305).

The spur for this crucial redefinition is the UN policy initiative to frame a new set of Sustainable Development Goals for the period 2016 to 2030 to replace the Millennium Development Goals following their expiry at the end of 2015. Griggs and his colleagues are contributing to this initiative through the transnational Sustainable Development Solutions Network Leadership Council (2013), which has produced an 'Action Agenda for Sustainable Development' that broadly accords with the 'nested' approach. For example, while economic development and ending extreme poverty remain key priorities, these are now to be achieved in ways that respect 'planetary boundaries' by ensuring environmentally sustainable production and consumption patterns and helping to stabilise the human population globally by mid-century. Improvements to agricultural systems, rural prosperity and urban quality of life are also to be keyed to enhanced environmental sustainability, which is now understood to include not only the protection of biodiversity and improved management of water and other natural resources, but also concerted climate change mitigation by means of arresting and reversing deforestation, along with a rapid shift to clean energy production 'for all'.

In the model presented in *Science*, Griggs and his colleagues synthesise the ten 'priority challenges' identified in the 'Action Agenda' into six over-arching goals – Thriving Lives and Livelihoods; Sustainable Food Security; Sustainable Water Security; Universal Clean Energy; Healthy and Productive Ecosystems; and Governance for Sustainable Societies – each of which cuts across the nested economic, social and environmental domains. This reframing of sustainability poses profound challenges for 'developed' as well as 'developing' nations, in addition to promising to redress the marginalisation of ecological considerations that has continued largely unabated, despite – perhaps even under the cover of – the proliferation of the rhetoric of sustainability. In the encompassing outer circle of Griggs's model, humans and nonhumans alike find themselves in the same 'nest', one that is at once constitutive of, and constituted by, the Earth's biosphere. The precise coordinates of this meta-nest are provided by the nine 'planetary boundaries' identified by Johan Rockstöm and colleagues at the Stockholm Resilience Centre, which define limits to climate change, biodiversity loss, changes to the nitrogen and phosphorus cycles, ozone depletion, ocean acidification, freshwater use, changes to land use (especially the conversion of wildlife habitat to agricultural or industrial purposes), chemical pollution, and atmospheric aerosol loading. Forms of economic and social development that transgress these boundaries, Griggs et al. warn, are liable to cause 'widespread, abrupt and possibly irreversible changes to basic Earth-system processes' (2013: 306).

At this point, however, a further possible pitfall of the sustainability agenda comes into view: namely that identified by Steve Mentz as a 'fantasy about stasis' (2012: 586). While Mentz's suspicion might hold true of some popular visions of sustainability, I do not believe that this is implicit in the notion of planetary boundaries, which presupposes the dynamism of biophysical systems whilst seeking to conserve the conditions under which the 'discordant harmonies' (Botkin 1990) discerned by post-equilibrium ecological science might continue to resound on Earth, in new variations and in new keys, into the future. What is to be sustained, on this understanding, is not a stable set of entities and relations, then, but the potential for ongoing or, in sites of pre-existing degradation and diminishment, renewed more-than-human flourishing. This in itself nonetheless implies a transformation of business-as-usual, not only for environmentally harmful industries, but also for prevailing environmental practice. For, in light of the new ecology, and in the grip of global warming, biodiversity conservation and ecological restoration can no longer consist in the endeavour to maintain species within, or return them to, their pre-existing geographical bounds: rather, both entail facilitating species' migration or hybridisation (Becker et al. 2013), as free-living plants and animals seek to forge their own pathways of survival into an uncertain future. Under today's intensifying conditions of heightened 'landscape fluidity', new models of environmental sustainability are needed, such as the seemingly oxymoronic notion of 'anticipatory restoration', as proposed by Adrian Manning and his colleagues (several of whom also co-authored 'Mind the Sustainability Gap') in their guest editorial to a 2008 issue of the *Journal of Biogeography*. This involves restoring the '*properties* of past functional ecosystems without attempting to create unattainable facsimiles of the past', not only in 're-wilded' zones, moreover, but also in mixed or 'cultural' landscapes, so long as, in these places, sustainable forms of 'commodity production' can conceivably co-exist with flourishing populations of free-living biota (Manning et al. 2008: 195). A further conceptual shift is required here too: for while modelling of climate change impacts continues to improve, we also need to get better at anticipating the unforeseeable. Recognising this element of incalculability necessitates the development of improvisational forms of 'adaptive governance' (Brunner and Lynch, 2010), as we seek to act responsibly under conditions of uncertainty. As I have argued elsewhere, this entails in turn honing our skills of creatively and compassionately 'dancing with disaster' (Rigby 2009, 2015b) in the face of the increasing frequency and intensity of weather-borne extreme events.

There is, nonetheless, a further flaw with conventional constructions of sustainability: one that is more fundamental than the 'fantasy of stasis' targeted by Mentz, and has evidently been carried over into the nested model of sustainable development. Here, as Stacy Alaimo has observed of the original Brundtland definition, '[n]ot only are the "generations" usually taken to be human, but the lively world is reduced to material for meeting their "needs"' (2012: 562). This unreflected anthropocentrism is indicative of the second problem identified by Fischer et al.: namely, a failure to 'reflect on foundational issues' and to 'confront potentially uncomfortable ethical questions' (2007: 623). It is in order to redress this shortcoming that they strongly advocate enhanced collaboration between natural and social scientists and humanities scholars, with a view to more effectively linking 'short-term policy actions with agreed longer-term sustainability targets' on the basis of 'critical analysis of foundational and longer-term issues (e.g. values, beliefs and motivations)' (Fischer et al. 2007: 623). This point is elaborated further in a co-authored Perspective piece in *Nature Climate Change* that desiderates the marginalisation of the environmental social sciences and humanities from current discussions of the 'human dimensions' of global climatic and environmental change (Castree et al. 2014). Conducted almost exclusively by natural and quantitative social scientists, existing research in this area 'offers little or no sense of humans as diverse, interpretive creatures who frequently disagree about values, means and ends; and there is no mention of power, violence, inequality and the perennial desire of some people to replace one socio-environmental regime with another' (Castree et al. 2014: 765). Occluding cultural, philosophical and socio-political differences in perspective, the sustainability agenda universalises a historically specific view of other-than-human 'nature' as a store-house of resources and provider of services for Earth's sovereign species. While this view might have counterparts in some non-Western civilisations, it enters the discourse of sustainability from a distinctively modern Western line of thinking. As decades of research in the environmental humanities have clearly demonstrated, this can be traced back to certain Greco-Roman and biblical notions of human exceptionalism, but became consolidated in that project of human mastery first formulated as such in the context of the scientific revolution (a project in which anthropocentrism was historically correlated also with androcentrism, as Theodor Adorno and Max Horkheimer noted in their *Dialectic of Enlightenment* of 1944 (1979: 3), and has since been examined in more depth and detail by ecofeminist scholars such as Carolyn Merchant (1980) and Val Plumwood (1993)). In order to move from 'paradigms of conquest to paradigms of connectivity', as Fischer et al. propose (2007: 623), or, in Plumwood's formulation of this shift, to disavow the Cartesian quest to extend the 'empire of man over mere things' in favour of negotiating 'life membership in an ecological community of kindred beings' (2009: 119, 121), a more thoroughgoing reconceptualisation of sustainability is required: one in which the current and future generations, whose needs are to be met, are understood to be more-than-human.

As already noted, consideration of more-than-human flourishing is given a higher priority in the nested model of sustainability to the extent that biodiversity loss constitutes one of the planetary boundaries that must not be transgressed (along with climate change, which is set to compound existing pressures on wildlife habitat and dramatically escalate the extinction rate). However, in the absence of an explicit affirmation of ethical regard for other-than-human beings in their own right, the tacit assumption here is that biodiversity loss should be limited primarily in order to protect human interests in the medium to long term. Some conservation biologists might well be among those who see the protection of biodiversity (including genetic diversity within species) as an end in itself, considering that while extinction is intrinsic to evolution, it is fundamentally unethical for one species, which is possessed of the cognitive capacity and moral discernment to do otherwise, to be condemning so many others to oblivion at the current calamitous rate. But in societies that remain highly 'anthroparchal' – characterised, that is, by systematic forms of human domination, exploitation and marginalisation of nonhuman others (Cudworth 2005: 63–70) – it is rather unsurprising that the value of biodiversity should commonly be framed primarily in terms of its human benefits. Yet, as Freya Mathews (2013 and 2016) has demonstrated, the anthropocentric case for biodiversity conservation is not only ethically questionable; it is also ultimately unconvincing.

Two of the most frequently cited grounds for biodiversity conservation are particularly flimsy. Firstly, the idea that we should save other species (generally of the charismatic kind) so that our grandchildren can have contact with them is readily countered by the argument that since people only miss what they have known, future generations are unlikely to care much about species that had disappeared before they were born, especially as they are likely to live on as simulacra (which is the only way that most children have contact with them today anyway, give or take the occasional zoo visit). Secondly, the claim that people need contact with nonhuman others and more-than-human places for their psychophysical health provides reasonable grounds for pet ownership, farm visits, country rambles, bush walks and the provision of urban parks and gardens; but it does nothing for the protection of free-living species in far-flung locations, where little human contact is feasible, or probably even desirable, on a regular basis. A far sturdier anthropocentric case for biodiversity protection can be made on the basis of its role in the provision of 'ecosystem services', and it is in these terms that it is framed in the draft Sustainable Development Goals. Yet, as Mathews observes, future technological advances in biomimicry, such as are already underway in the development of robotic pollinators to compensate for the decline in honeybee populations, hold out the possibility that our fledgling high-tech civilisation could plausibly fly the biospheric nest in which it hatched, enabling future generations of humans to get along just fine with a radically reduced suite of other species, retained either for their ornamental, companionate or instrumental value.[2]

A further weakness in the anthropocentric ethic underpinning conventional framings of biodiversity conservation that Mathews identifies is that it only triggers intervention at the point of endangerment, thereby tending towards an 'ecology of last things' (see Joshua Schuster's essay in this collection). This is actually an oxymoron, as flourishing ecosystems rely not only on dynamic interrelations among different species, but also on the relative abundance of each species (with far fewer top predators, for example, constituting a viable population by comparison with herbivores and invertebrates). Importantly, population size is also a key factor in maintaining genetic diversity, and hence the capacity of species to adapt to changing conditions. Mathews therefore proposes instead an ethic of 'bio-proportionality', grounded upon a generalised respect for living things in themselves, rather than simply as service providers for humans. The express goal of bio-proportionality is the optimisation of populations of all members of those multi-species collectives (frequently including humans) whose dynamic interrelations engender ecosystemic flourishing.

In the era of the Anthropocene, on a planet increasingly given over to servicing exclusively human domiciles (however inequitably), the principle of bio-proportionality enjoins an ethic of bio-inclusive hospitality: the imperative, that is, to make space on 'our' Earth for the domiciling of 'otherkind' (an ecotheological coinage that felicitously conjoins recognition of alterity and kinship in 'humankind's' relations with other creatures).[3] Bio-proportional hospitality falls short of Jacques Derrida's definition of the radical categorical imperative of hospitality. This enjoins an unconditional welcome to any and every 'arrivant': 'whoever or whatever turns up', that is, 'whether or not the new arrival is the citizen of another country, a human, animal, or divine creature, a living or dead thing, male or female' (Derrida and Dufourmantelle 2000: 75, 77), prior to any identification, without any expectation of reciprocity, and beyond any possible calculation of collective wellbeing. In practice, however, hospitality towards the 'arrivant' is inevitably always qualified by one's other duties of care, as Derrida reminds us with the tale of the biblical patriarch, Lot, himself a non-native inhabitant of Sodom, who offered up his own virgin daughters in place of his angelic guests to the Sodomites who wished to 'penetrate' them. An ecopolitical analogue of this might be the actions of those legislatures that have sought to protect wildlife habitat, questionably construed as 'wilderness', at the cost of expelling indigenous peoples. By contrast, Mathews's bio-proportionality ethic envisages multi-species collectives, in which humans might well play a critical role in 'caring for country', to use an Aboriginal English expression, as is already the case across large swathes of central and northern Australia, where Native Title holders, drawing on their traditional ecological knowledge and skills in conjunction with a selective use of contemporary science and technology, are engaged in vital conservation and (increasingly, anticipatory) restoration efforts (Altman and Kerins 2012).[4]

Mathews also foresees a vital place for those techno-scientific advances (for instance, in nutrition, housing, water use and energy production) that could help to relieve the human pressure on the biosphere as we transition to a new, 'ecological civilisation'.[5] In addition to constraining human population growth and consumption levels through democratically-instituted forms of sustainable development that promote social equity and inclusion within the framework of a bio-inclusive ethic of more-than-human flourishing, bio-proportional hospitality would mandate the opening up of migration corridors for species unhoused by climate change, as well as making provision for otherkind in the face of increasingly frequent and intense extremes. Such practices instantiate what I have elsewhere termed 'ecstatic hospitality' (Rigby 2008), modelled biblically in the figure of Noah's Ark, in which refuge is offered by a host whose own home too is unmoored and liable to be lost.

To frame bio-proportionality in terms of hospitality is to wager on altruism. Personally, I think highly enough of human potential (despite ample contrary evidence) to consider this wager worthwhile. But I am also enough of a realist (and, for that matter, an epicurean) to share the view of Kate Soper and her colleagues that, at least among the more pampered citizens of the 'developed' world, the pleasure principle is likely to provide a more powerful motivation for the kind of socio-ecological transformation entailed in the bio-inclusive practice of sustainability. To incline people towards 'deep sustainability', though, might require a 'deeper' kind of pleasure than those forms of 'alternative hedonism' identified thus far under this rubric, such as 'slow food', self-provisioning, cycling and sensual immersion in 'wild' places (Soper et al. 2009). In Mathews's analysis, it necessitates nothing less than the 'transvaluation of desires' (2010: 3), entailing a fundamental re-orientation towards materiality *per se*.

Within the limits of this chapter, it is not possible to expand upon the onto-epistemological underpinnings of this proposed re-orientation, as explicated by Mathews in her monograph *For Love of Matter* (2003) and explored further in *Reinhabiting Reality* (2005).[6] Put (far too) simply, though, Mathews's 'contemporary panpsychism', like Plumwood's 'philosophical animism' (2009) and other variants of 'new materialism' (e.g. Coole and Frost 2010), challenges the prevalent view of matter as passive, mute and mindlessly mechanistic that came to prominence with Cartesian dualism and Newtonian atomism. The inadequacy of this view was already becoming apparent to those physicists, such as Werner Heisenberg and Niels Bohr, who began exploring the weird and wonderful world of quantum mechanics in the 1920s, and it is now being challenged more widely by contemporary physicists such as Karen Barad (2007). In the meantime, though, reductive materialism had become rooted in modern Western culture, where it found a neat fit with consumer capitalism. Stripped of creative agency, communicative capacity and ethical considerability, the realm of 'mere matter' becomes available to be mined, manipulated and disposed of in whatever way those humans with the buying power to do so think fit.

In Mathews's analysis, this impoverished view of reality also profoundly impoverishes human existence, no matter how rich in stuff it has made some of us, in that it tends to limit our potential for self-actualisation in and through our intersubjectival relations with others to interactions with fellow humans (interactions that are themselves increasingly semiotically diminished by being reduced to words on a screen), albeit possibly supplemented by animal companions and/or supernatural deities. The reductively materialist metaphysics that haunts modern Western culture in turn feeds the hunger for ever new, ever disposable consumer trinkets precisely because, perceived as 'mere things', valued not even principally for their utility so much as for the social identities they allow their owners to embrace and display, they forever fail to satisfy our deeper longing to participate in an inherently meaningful more-than-human world. While Mathews has presented carefully reasoned arguments for her alternative, monist metaphysics for the benefit of fellow philosophers, her version of the cultural work of deep sustainability is more practical than theoretical. In order to ween ourselves from reductively materialist and socio-ecologically unsustainable forms of commodity fetishism, Mathews recommends the cultivation of practices that afford the deeper pleasures of interactive self-actualisation, or co-becoming, through experiences of intersubjective encounter, communicative interchange and, potentially, synergistic co-creation with more-than-human others and those places in which we might meet with them. Mathews terms such interactions 'ontopoetic' (2009), and in the remainder of this chapter I consider some examples of literary works that are conducive to this ontopoetic transvaluation of desire in ways that are also consistent a bio-inclusive practice of hospitality.

Refiguring the nest: the ecopoetics of deep sustainability from Clare to Morley

If, as suggested previously, the bird's nest is taken to be both literally and figuratively emblematic of the life-sustaining work of contributing to the other's flourishing, then there is surely no better English-language writer to consider in this connection than John Clare, who penned some one hundred poems dedicated to birds and their nests. Clare, along with several other Romantic writers and philosophers, has attracted a good deal of ecocritical attention following the publication of Jonathan Bate's landmark study of 'romantic ecology' in 1991; and in his later monograph, *The Song of the Earth*, Bate homed in on Clare's nest poems in particular as exemplary of an ecopoetics of dwelling. Yet, as Richard Kerridge observes in his discussion of 'Green Pleasures' (2009), Romanticism occupies an ambivalent position in relation to sustainability. In the influential analysis of Colin Campbell (1987), the Romantic celebration of the human capacity to imagine possibilities that render everyday realities disappointing by comparison nurtured an ethos of inchoate longing that became the cradle of insatiable consumerist desire. Campbell stressed, however, that this constitutes a historical irony, as most Romantics took a more or less explicitly critical view of the growing commercialism of their day: 'Getting and spending we lay waste our powers', proclaimed Wordsworth in 'The World is Too Much with Us', as Kerridge recalls (2009: 142). In Kerridge's view, this is an irony that 'might conceivably cut both ways. If Romanticism provides the structure of desire that motivates consumerism, then Romanticism remains powerfully latent in contemporary culture: there to be renewed in non-consumerist forms' and capable, perhaps, of providing a 'bridge between pre-industrial and the post-industrial sensibility' (Kerridge 2009: 146, 147): a possibility explored further by Kate Soper in her discussion of 'avant-garde nostalgia and hedonist renewal' (2011). To this I would add that Romanticism is in any case an inchoate historical phenomenon, such that all summary characterisations, including Campbell's, 'imply a coherence ... that close inspection calls into question' (Day 1996: 5).[7]

Within the highly heterogeneous field of British Romantic literature, Clare definitely did not share the fetish for inchoate longing, preferring instead precisely that mode of close observation of the other-than-human inhabitants of his rural environs in all their material particularity and in a spirit of non-appropriative empathetic attentiveness which Mathews identifies as a critical first step towards the ontopoetic transvaluation of desire (2010: 3–4). Take, for instance, 'The Nightingales Nest', a poem that cuts its figure against a long-standing literary tradition of putting the nightingale, and above all his (or, as is more often the case in poetry, if not in nature, 'her') nocturnal song, to symbolic purposes. One of the most prevalent poetic uses of the nightingale since classical times is as a figure for the joy and anguish of amorous love, and it is this anthropomorphising trope that Clare initially invokes:

> Her wings would tremble in her ecstasy
> And feathers stand on end as 'twere with joy
> And mouth wide open to release her heart
> Of its outsobbing songs. (Clare 2004: 168)

While the charge of anthropomorphism, as Plumwood argues, is all too often deployed as a 'policeman for reductive materialism' (2009: 126), patrolling the boundaries of human–nonhuman hyperseparation, the conventionalised projection of human-referenced attributes onto a nonhuman other is also problematic if it blocks recognition of alterity and singularity. Clare is evidently mindful of this risk, as he cites the conventional figuration of the nightingale's song precisely in order to depart from it: firstly, this bird is said to be singing all day, not all night; secondly, the use of the conditional, 'as 'twere', signals a note of uncertainty about what this bird might actually be feeling. In the following lines, moreover, her apparent enjoyment of the 'happiest part / Of summer's fame' is referenced to the poet's own 'happy fancies'. Such culturally conditioned imaginings are counter-posed to the corporeal as well as mental effort entailed in trying to apprehend this particular bird in her own sphere of existence and world of signification: her distinctively avian *Umwelt*, in Jakob von Uexküll's terminology (2010).[8] In so doing, however, Clare also draws attention to the creatureliness shared by bird and human, even while stressing the necessity of getting out of his or her comfort zone for the would-be birdwatcher, who is depicted '[c]reeping on hands and knees through matted thorn / To find her nest and see her feed her young.' Clare also hints at the conjunction of human–animal similarity and alterity in the preceding line, which foregrounds the appetitive aspect of this quest: 'There have I hunted like a very boy' (2004: 168). This simultaneously links his current practice to, and distinguishes it from, the nest-raiding of 'rude boys' (such as Clare himself had once been) – an activity now perceived as morally dubious, being motivated not by subsistence requirements but by the perhaps distinctively human thrill of non-nutritive collection: one that the nightingale has learned to foil by building her 'secret' nest 'where rude boys never think to look' (Clare 2004: 169).

As an adult, the speaker has evidently discovered where to look; but the desire that propels his search is no longer appropriative, but empathetic, and, in Mathews's terms, incipiently erotic. Eros, in this view, refers to the desire not to lustfully possess the other, but to connect with them in such a way as to respect their alterity, whilst seeking a mutual flourishing. The transition away from the 'brute-striving of appetite' to an 'awakened reaching-out' (Mathews 2003: 150) is signalled in the shift from the speaker's initial injunction to his interlocutor in the opening lines to 'softly rove' and 'Hush!', which serves the ambivalent purpose of enabling them to sneak up on the nightingale, more for their benefit than for hers, to the later exhortation to resist the temptation to 'trample' on the brambles to access her nest, mindful that 'our presence doth retard / Her joys' (Clare 2004: 170). Advancing a bio-inclusive ethic of respect for the dwelling-places of otherkind, the speaker urges his companion (and the poet thereby his readers) to leave 'her home ... as we found it: safety's guard / Of pathless solitudes shall keep it still' (Clare 2004: 170). This shift is prompted by an empathetic attentiveness to the way in which the bird is reading and responding to the human intruders, the cessation of her song, alarm call ('a plaintive note of danger' (Clare 2004: 170)) and anxious movements being legible as signs of fear on the basis of the shared creatureliness that subtends the acknowledged differences between human and avian semiospheres. This call to compassionate self-restraint is nonetheless followed by a rapturous address to the avian songstress, hailing the 'melody' that 'seems hid in every flower / That blossoms near thy home', and proceeding to a lingering description of her 'curious' and elusive nest (Clare 2004: 170):

> no other bird
> Uses such loose materials or weaves
> Its dwelling in such spots – dead oaken leaves
> Are placed without and velvet moss within
> And little scraps of grass and, scant and spare,
> What scarcely seem materials, down and hair. (Clare 2004: 170)

While the poem concludes with a reiteration of the call for restraint in order to protect the bird's hidden nesting-place with its five 'curious eggs', altruistic concern for the other's flourishing has now been joined by erotic enjoyment of non-appropriative contact with a kindred being, along with her glorious song and distinctive *Umwelt*.

In his discussion of 'The Yellowhammer's Nest', Washington remarks that the titular nest is portrayed by Clare as 'a unique composition, expertly crafted by this individual bird within a particular ecology' (Washington 2014: 668). Such nests appear as the product of a creative agency that exists on a continuum with that of the poet in weaving his work of words, which in turn bears witness to, and in that sense upholds, the bird's handiwork. Clare felt that his own literary productivity was actually gifted to him by the multiple agencies of his more-than-human environs, maintaining that he 'found the poems in the fields' (cited in Bate 2003: 15). Like Patrick Bresnihan, I do not believe that this should be taken as a mere figure of 'poetic sentiment'. Rather, it was indicative of Clare's recognition that his poetry was materially co-constituted by 'the force of the world acting on him', it arose from his embodied encounters with diverse others in a particular space and time, and as he attended to 'the way self and world were revealed, or achieved, through ongoing relations' (Bresnihan 2013: 80).

While Clare indubitably draws on pastoral tropes and traditions in much of his verse, he does not depict the multi-species collectives that nourished his writing as entirely harmonious. As 'one of the great poets to chronicle the daily lives of animals, their sounds and shapes, their habits and habitats, their wonder and welfare', Clare also records not only their 'sorrows and sufferings at the hands of humans' (Washington 2014: 665), but also, as in the case of the yellowhammer pair whose nest is raided by a peckish snake, those that arise in the normal course of creaturely existence. The changes that he perceived to be unfolding in association with the enclosure of erstwhile common land, though, exposed other creatures to a whole new regime of human domination, as well as undermining the subsistence needs of the rural poor. Clare's concern about this wider threat to the domiciles of free-living animals doubtless informs the anxiety that the speaker betrays in 'The Nightingale's Nest' regarding the invasiveness of his own birdwatching activities. This link is implicit in the affirmation, 'We will not plunder music of her dower / Nor turn this spot of happiness to thrall' (Clare 2004: 170), *thrall* being the word used to characterise the relationship of mastery and possession instituted by enclosure in Clare's protest poem, 'The Mores': 'These paths are stopped – the rude philistine's thrall / Is laid upon them and destroyed them all' (2004: 91). It is perhaps also this wider context of socio-ecological change that motivates the move, in the concluding lines of 'The Nightingale's Nest', out of the mode of loving attentiveness to the particular – which, as Clare surely knows, can never be captured in its concrete singularity in any work of words, no matter how skilfully crafted – into a more abstract and generalising register: 'So here we'll leave them, still unknown to wrong, / As the old woodland's legacy of song' (2004: 171). This 'legacy' can be read literally as the perpetuation of birdsong down the avian generations, facilitated by the protection of their nesting-places. But it might also be understood figuratively, with reference to something more like a *genius loci*, such as that embodied by the speaking brook in 'The Lamentations of Round-Oak Waters'. As such it refers to a circumambient sentience that 'still' inheres in the inter-specific co-becoming of the commons, but that is being silenced by the conversion of this 'animated, sensible landscape' (Irvine and Gorji 2013: 123)[9] into a mere storehouse of 'natural resources': land to be rid of 'pests', such as the moles 'hung .... as traitors', as Clare puts 'Remembrances' (2004: 134), and rendered 'productive', no longer of vibrant multi-species collectives, but of cash crops and, thereby, taxable income for private land-owners.

As Bresnihan observes (2013: 79), Clare offers a telling image of the new mentality associated with enclosure in one of his earliest poems, 'A Ramble'. Here, the speaker's enchantment with 'every trifle nature's bosom wears' is contrasted with the indifference of the 'heedless passenger', who:

> Soodles me by, an animated post,
> And ne'er so much as turns his head to look
> But stalks along as though his eyes were blinded
> And as if the witching face of nature
> Held but now a dark unmeaning blank. (Clare 2004: 8)

Read in conjunction with Mathews's critique of reductive materialism, these lines disclose how the simultaneously de-animating and, as suggested by 'stalks', predatory mindset of industrial modernity-in-the-making also devitalises human subjectivity. A world stripped of its inherent meaningfulness, reduced to a passive screen for human projections and a means for human ends, is one in which the self-proclaimed sovereign subject too is psychically diminished. Reduced to an 'animated post', she too is liable to be instrumentalised as part of the support structure for a system in which anyone, as well as anything, can be reduced to what Martin Heidegger in 'The Question Concerning Technology' (1993) aptly termed 'standing reserve' (analogous, that is, to trees defined as timber, left standing only to be felled).

Arrested in his self-actualisation as a psychophysical being in communion with an agentic and communicative more-than-human world, the possessive individual of the new enclosed order, where 'Fence meets fence in owners' little bounds', finds themselves also 'imprisoned, ill at ease', as Clare puts it in 'The Mores' (2004: 90). The dis-ease occasioned by this cut-off condition (for which compensation would later be found in the increasingly frenetic shopping of commodity fetishism) contrasts with the psychophysical nourishment afforded by the interconnectedness of co-becoming, such as Clare celebrated, for example, in his fond recollection of co-habitation with the big old tree, whose felling he laments in 'The Fallen Elm' (2004: 141–3). Hailed as a 'friend not inanimate', who 'murmured in our chimney top / The sweetest anthem autumn ever made', this beloved tree is said to have provided 'comfort to our heart's desire', summer shade for children's play and a nesting-place for the mavis (a thrush, whose highly musical song would no doubt have occasioned further delight following the birds' return from their annual migration to breed). By contrast with the animate character of this hospitable arboreal companion, the landlord who had it felled is implicitly classed among those soulless humans who (recalling the 'animated post' of 'A Ramble') are figured as 'stocks and stones ... many formed of flesh and bones'. Here too, a further dimension of the mindset of the new order emerges in Clare's ironic references to its rhetoric of 'freedom', whereby the legal freedom of property owners to dispose of their possessions however they see fit is shown to be eroding the liberty, livelihood and hence life chances of the rest. Thus, for instance, when:

> The common heath – became the spoiler's prey:
> The rabbit had not where to make his den
> And labour's only cow was drove away.
> ...
> Such was thy ruin, music-making elm:
> The rights of freedom was to injure thine.
> As thou wert served, so would they overwhelm
> In freedom's name the little that was mine. (Clare 2004: 143)

In addition to modelling ecopoetically the cultural work of deep sustainability, then, Clare's verse provides a diagnosis of the roots, at once socio-economic and psycho-social, of the potentially ecocidal trajectory of industrial modernity in the failure to cultivate delight in, and respect for, the domiciles and lifeways of more-than-human others and the communicative matrices co-created by their vital interactions. 'To a Fallen Elm', in which the impact of the enclosed order cuts closest to home for Clare's speaker, has acquired a new salience, moreover, in light of the loss of almost all elm trees throughout Britain, as well as much of Europe, Canada and New Zealand. 'Dutch Elm disease' was first identified in Holland in the 1920s, but the most recent and considerably more pathogenic strain of the beetle-borne fungus that affects elms, and to which the iconic English elm is particularly vulnerable, entered Britain only in the 1960s, evidently on a shipment of timber from Canada (Gibbs et al. 1994): Britain's elms, then, have now fallen victim *en masse* to the transnational trade in 'natural resources', a key element in the unsustainable 'Great Acceleration' of the industrial order of the Anthropocene that Clare saw taking hold in Helpston in the 1820s.[10]

On the eve of a new outbreak of this disease that was reported to be threatening Britain's remaining English elms in 2010 (Seddon 2010), however, some of the trees that had died previously were afforded a kind of afterlife in an ecopoetic art installation undertaken by David Morley as part of a Slow Art project initiated by Chrysalis Arts on the grounds of the Bolton Abbey Estate in Yorkshire in 2008. Morley's indebtedness to Clare as an ecopoet is most evident in the tribute that he pays his Romantic predecessor in a book-length sonnet series entitled *The Gypsy and the Poet* (2013), which also honours the 'gypsy', Wisdom Smith, another frequenter of the disappearing commons with whom Clare is known to have conversed. Himself of Roma heritage, Morley is by training a conservation biologist, and committed to exploring how poetry might contribute materially to increasing the species diversity of a given habitat. In his contribution to the Slow Art installation, he attempted this by acquiring some elm planks that had been stored long enough for the offending fungus to depart, on separate pieces of which he carved a series of haiku, written in response to the remnant of ancient oak forest, Strid Wood, in which they were then sited on short poles. The carved words of these 'Ankle-High Haiku' were filled with potter's clay and consequently teemed with microbes from the human hands in which it had been moulded. These literally living words of clay were subsequently 'read', firstly, by algae attracted to the microbes, then by lichens attracted by the algae, which in turn lent the letters a greenish hue, rendering them more visible to any passing humans (especially children, in whose eyeline they are located), as well as attracting birds, who 'read' them in their own way as a source of suitable nesting material. Morley's ecopoetic experiment instantiates a synergistic practice of more-than-human co-creation by opening a space that invites other species to get in on the act as they appropriate his artwork to enlarge their habitat. Stumbling upon Morley's 'Ankle-High Haiku', possibly while waiting for their canine companions to add to the species diversity of the upcycled elm planks by pissing on them, human visitors to Strid Wood, meanwhile, are invited by these literally green words to attend more closely to their wooded environs. Coming upon 'Sussuration', for instance, they might read:

> The Academy
> Of Ancient Root systems is
> Open. Hush. Listen. (Morley 2014)[11]

Prompted to listen for the sound of wind in the leaves, the audible self-disclosure of the trees, human readers are encouraged also to consider, and ideally delight in, the material intelligence, creative agency and communicative capacity of the varied more-than-human others who have co-created the space in which they find themselves.

It was, as it happens, Morley's erstwhile colleague at Warwick University, Jonathan Bate, who launched the ecocritical re-evaluation of Clare and, in particular, his bird's nest poems to which Morley's Slow Poetry project also pays tribute, albeit more obliquely. In its interweaving of Heideggerian phenomenology, Adorno and Horkheimer's Marxist critique of the domination of nature within capitalist modernity, and Michel Serres's notion of a 'natural contract', Bate's take on Clare in *The Song of the Earth* was important in foregrounding the relationship between human psychophysical wellbeing and socio-ecological conditions. As I have argued elsewhere (Rigby 2004), however, I think that in his reception of Heidegger, Bate is lured into an anthropocentric over-valuation of the poetic word: while a poem might invoke and, in its musicality partially echo some of Earth's diverse more-than-human voices, to cast any work of merely human words as 'the song of the earth' (Bate 2000: 251) risks falling prey to a colonising kind of human self-aggrandisement. From a posthumanist material ecocritical perspective, then, it is important to stress that for all the tender protectiveness that Clare's poems evince towards birds and their nests, they themselves are, as it were, empty nests, inevitably failing to provide the space for the nurturance of flourishing more-than-human life that he so desperately, and ultimately despairingly, sought to safeguard. As an element in the wider discursive–material matrix in which humans intra-act with more-than-human others, however, literature that invites its readers to pay empathetic attention to the surprising lives and strange *Umwelten* of otherkind can potentially help to foster the bio-inclusive ethos of hospitality that necessarily undergirds a bioproportional model of sustainability, as well as awakening a desire to experience the kinds of ontopoetic encounter that such literature might invoke. But only 'potentially': for, to recall a motto of the Scholastics, 'whatever is received is received according to the mode of the receiver' (*quodquip recipitur ad modum recipientis recipitur*; Bretzle 2013: 200), which is in turn dependent upon a host of extra-literary material–discursive factors (cultural, social, political, economic, geographic, institutional, personal etc.).

The same is true, of course, of the human reception of Morley's 'Ankle-High Haiku'. As a work of ecopoetics, however, the Slow Poetry trail marks a necessary movement beyond both the cultural-historical and aesthetic limits of Romantic nature poetry. At a time of escalating anthropogenic extinctions, Clare's ethic of letting be is insufficient: the restoration of bioproportionality demands that humans actively create affordances for the flourishing of other species, whose habitat, like that of the woodland birds invited to avail themselves of the lichen on Morley's 'Ankle-High Haiku', is being anthropogenically eroded.[12] Moreover, at a time when ever more people, and especially children, are being lured into simulacral worlds of more or less exclusively human construction, ecopoetic projects that draw their recipients outdoors have a better chance than mere words on a page of fostering a taste for the alternative hedonism afforded by spending time with free-living plants and animals and the lively, communicative and (if we follow Mathews's panpsychist premise) sentient places in which they might be met. Making material provision for other-than-human dwelling through creative practices of bio-inclusive hospitality, ecopoetics beyond-the-page simultaneously works towards the transvaluation of human desires by opening recipients to the deep pleasure of ontopoetic encounter.

Clearly there is much that needs to be done in a range of ways and contexts, and with varying degrees of urgency and difficulty, to bridge the sustainability gap. Literature, especially that which entails and discloses intra-active processes of multi-species co-creation, can play a part in this by contributing to that shift in underlying attitudes, assumptions, values and desires which would be conducive to the safeguarding of planetary boundaries in the interests of the renewed flourishing of Earth's diverse more-than-human life.

Notes

1 Plumwood's move into literary territory in this essay was encouraged by the invitation to present a keynote lecture at the 2002 conference of the UK Association for the Study of Literature and Environment: the essay is based on that plenary presentation, and enriched by the discussions it occasioned.
2 See also Conniff (2012) for a succinct discussion of a number of other lines of critique of the concept of 'ecosystem services' and, in particular, attempts to price them.
3 See e.g. Nash, who refers to 'all creatures, human and otherkind' (1996: 9). More recently, Anne Elvey has defined this term more inclusively to include 'both those we understand as living (e.g., fleas, whales, and eucalypts) and those we understand otherwise (e.g., glaciers, sand, and air)' (2014: 36).
4 'Caring for country' should not be confused with Western ecofeminist 'ethics of care'. It has a foundation in traditional ecological knowledge ('Law'), rather

than sentiment (although Indigenous Australians do evince a high degree of affective attachment to their ancestral homelands) (Rose et al. 2002). Presupposing more-than-human agency, communicative capacity and human–nonhuman connectivity and kinship, 'caring for country' entails something rather more like the considered practice of intra-active material–discursive interrelationship among diverse human and nonhuman actants envisaged by Adeline Johns-Putra (2013) in her new materialist model of environmental care. The (thus far) most extensive historical study that amply evidences the socio-ecological efficacy of Aboriginal land management in sustaining very high levels of biodiversity over extremely long time periods is Gammage (2011). It should be stressed that this case should not be assumed to be universally representative of indigenous culture *per se*, not does it preclude the possibility that the distant ancestors of Australia's First Nations peoples might have inadvertently contributed to the extinction of the megafauna that they encountered on this continent some 40–60,000 years ago, many of which nonetheless co-existed with humans for at least another 30,000 years, succumbing only after the climate changed from cold-dry to warm-dry and water became scarcer at the end of last glacial maximum (Musser 2014).

5 'Ecological civilisation' is a Chinese concept with roots in Taoist philosophy that Mathews has been studying for several years, in collaboration with colleagues in China. Mathews currently holds the position of Adjunct Professor of Eco-Civilisation Studies at Monash University's Institute of Sustainability.

6 For an overview, see Rigby (2006).

7 On Romanticism and ecocriticism, see Rigby (2014).

8 Clare's attention to species-specific *Umwelten* is also discussed by Washington (2014: 666) and examined in the wider context of Romantic-era precursors to biosemiotics in Rigby (2015a).

9 In this connection, Irvine and Gorji cite from a letter Clare wrote in connection with his relocation from Helpston to Northborough: 'the very molehills on the heath and the trees in the hedgerow seem bidding me farewell' (2013: 123).

10 Evidence for the Great Acceleration is provided in Steffen (2004) in a series of graphs charting changes in human activities and correlating environmental impacts along j-curves, all of which take off during or from the 1950s.

11 This is also discussed in the series of Slow Poetry videos available at www2.warwick.ac.uk/newsandevents/audio/more/slowpoetry.

12 Morley's contribution to the Slow Art project also involved the commissioning of a series of bird-boxes, on each of which he inscribed a poem that he had written in response to the morphology, habits and vocalisation of the particular species for which the box was likely to be most attractive. Dubbed 'Bard Boxes', these were then sited in appropriate places to afford additional nesting opportunities (Morley 2014).

References

Adorno, Theodor and Max Horkheimer [1944] 1979. *Dialectic of Enlightenment*. Trans. John Cumming. London: Verso.

Alaimo, Stacy 2012. 'Sustainable This, Sustainable That: New Materialisms, Posthumanism, and Unknown Futures', *PMLA* 127 (3): 558–64.

Altman, Jon and Seán Kerins (eds) 2012. *People on Country: Vital Landscapes, Indigenous Futures*. Sydney: Federation Press.

Barad, Karen 2007. *Meeting the Universe Halfway: Quantum Physics and the Entanglement of Matter and Meaning*. Durham, NC: Duke University Press.

Bate, Jonathan 1991. *Romantic Ecology: Wordsworth and the Environmental Tradition*. London and New York: Routledge.

Bate, Jonathan 2000. *The Song of the Earth*. Boston: Harvard University Press.

Bate, Jonathan 2003. *John Clare. A Biography*. London: Picador.

Becker, Matthias et al. 2013. 'Hybridization May Facilitate *In Situ* Survival of Endemic Specie through Periods of Climate Change', *Nature Climate Change* 3: 1039–43. www.nature.com/nclimate/journal/v3/n12/full/nclimate2027.html. Accessed 15 February 2017.

Botkin, Daniel 1990. *Discordant Harmonies: A New Ecology for the Twenty-First Century*. Oxford: Oxford University Press.

Bresnihan, Patrick 2013. 'John Clare and the Manifold Commons', *Environmental Humanities* 3 (1): 71–91.

Bretzle, James T. 2013. *Consecrated Phrases: A Latin Theological Dictionary*. Collegeville, MN: Liturgical Press.

Brunner, Ronald B. and Amanda H. Lynch 2010. *Adaptive Governance and Climate Change*. Chicago: University of Chicago Press.

Campbell, Colin 1987. *The Romantic Ethic and the Spirit of Modern Consumption*. Oxford: Blackwell Publishing.

Castree, Noel et al. 2014. 'Changing the Intellectual Climate', *Nature Climate Change* 4: 763–8.

Clare, John 2004. *Selected Poems*. Ed. Jonathan Bate. London: Faber & Faber.

Conniff, Richard 2012. 'What's Wrong with Putting a Price on Nature?', *Yale Environment 360*, 18 October. http://e360.yale.edu/feature/ecosystem_services_whats_wrong_with_putting_a_price_on_nature/2583/. Accessed 15 February 2017.

Coole, Diana and Samantha Frost (eds) 2010. *New Materialisms: Ontology, Agency, and Politics*. Durham, NC: Duke University Press.

Cudworth, Erica 2005. *Developing Ecofeminist Theory: The Complexity of Difference*. London: Palgrave Macmillan.

Cuomo, Christine J. 1998. *Feminism and Ecological Communities: An Ethic of Flourishing* London: Routledge.

Day, Aidan 1996. *Romanticism*. London: Routledge.

Derrida, Jacques and Anne Dufourmantelle 2000. *Of Hospitality: Anne Dufourmantelle invites Jacques Derrida to Respond*. Trans. Rachel Bowlby. Stanford, CA: Stanford University Press.

Elvey, Anne 2014. '"Who is the Greatest?" Reading Luke 22.24–27 Ecologically', *Ecotheology* 1: 35–48.

Eichler, Margrit 1999. 'Sustainability from a Feminist Sociological Perspective: A Framework for Disciplinary Reorientation'. In *Sustainability and the Social Sciences: A Cross-disciplinary Approach to Integrating Environmental Considerations into Theoretical Reorientation*. Ed. Egon Becker and Thomas Jahn. London: Zed Books: 182–206.

Fischer, Joern et al. 2007. 'Mind the Sustainability Gap', *Trends in Ecology and Evolution* 22 (12): 621–4.
Gammage, Bill 2011. *The Biggest Estate on Earth: How Aborigines Made Australia*. Sydney: Allen & Unwin.
Gibbs, John, Clive Brasier and Joan Webb 1994. 'Dutch Elm Disease in Britain', Forestry Authority Research Note 252. www.forestry.gov.uk/pdf/RIN252.pdf/$file/RIN252.pdf. Accessed 15 February 2017.
Griggs, Dave et al. 2013. 'Sustainable Development Goals for People and Planet', Comment in *Science* 495 (21 March): 305–7.
Heidegger, Martin 1993. 'The Question Concerning Technology' [1953]. Trans. William Lovitt. In *Basic Writings*. Ed. David Farrell Krell. San Francisco: HarperSanFrancisco, 307–42.
Irvine, Richard and Mina Gorji 2013. 'Clare in the Anthropocene', *Cambridge Anthropology* 31 (1): 119–32.
Johns-Putra, Adeline 2013. 'Environmental Care Ethics: Notes toward a New Materialist Critique', *Symplokē* 21 (1–2): 125–35.
Kerridge, Richard 2009. 'Green Pleasures'. In *The Politics and Pleasures of Consuming Differently*. Ed. Kate Soper, Martin Ryle and Lyn Thomas. Houndmills: Palgrave Macmillan, 130–53.
Manning, Adrian D. et al. 2008. 'Landscape Fluidity: A Unifying Perspective for Understanding and Adapting to Global Change', *Journal of Biogeography* 35 (2): 193–9.
Mathews, Freya 2003. *For Love of Matter: A Contemporary Panpsychism*. New York: State University of New York Press.
Mathews, Freya 2005. *Reinhabiting Reality: Towards a Recovery of Culture*. New York: State University of New York Press.
Mathews, Freya 2009. 'Introduction: Invitation to Ontopoetics', *Philosophy Activism Nature* 6: 1–6.
Mathews, Freya 2010. 'On Desiring Nature', *Indian Journal of Ecocriticism* 3: 1–9.
Mathews, Freya 2013. 'Is an Ethic of Biodiversity Enough?' *The Conversation*, 7 February, https://theconversation.com/is-an-ethic-of-biodiversity-enough-11425. Accessed 15 February 2017.
Mathews, Freya 2016. 'From Biodiversity-Based Conservation to an Ethic of Bioproportionality', *Biological Conservation* 200: 140–8.
Mentz, Steve 2012. 'After Sustainability', *PMLA* 127 (3): 586–92.
Merchant, Carolyn 1980. *The Death of Nature*. London: Wildwood.
Morley, David 2013. *The Gypsy and the Poet*. Manchester: Carcanet.
Morley, David 2014. 'Natural Magic'. Lecture presented to the Literary and Cultural Research Network, Monash University, 28 July.
Musser, Anne 2014. 'Megafauna Extinction Theories: Patterns of Extinction'. National Museum of Australia. 30 October. http://australianmuseum.net.au/megafauna-extinction-theories-patterns-of-extinction. Accessed 15 February 2017.
Nash, James A. 1996. 'Toward the Ecological Reformation of Christianity', *Interpretation: A Journal of Bible and Theology* 50 (1): 5–15.
Plumwood, Val 1993. *Feminism and the Mastery of Nature*. London: Routledge.
Plumwood, Val 2006. 'Deep Sustainability as Cultural Work'. Unpublished contribution to Sustainability workshop, ANU, June. Accessible on 'Remembering Val

Plumwood: A Memorial Site to Honour the Life and Work of Val Plumwood' http://valplumwood.wordpress.com/category/sustainability/. Accessed 15 February 2017.

Plumwood, Val 2007. 'Journey to the Heart of Stone'. In *Culture, Creativity and Environment: New Environmentalist Criticism*. Ed. F. Becket and T. Gifford. Amsterdam: Rodopi, 17–36.

Plumwood, Val 2009. 'Nature in the Active Voice', *Australian Humanities Review* 46: 113–29.

Rigby, Kate 2004. 'Earth, World, Text: On the (Im)possibility of Ecopoiesis', *New Literary History* 35 (3): 427–42.

Rigby, Kate 2006. 'Minding (about) Matter: On the Eros and Anguish of Earthly Encounter', *Australian Humanities Review* 38 (April). www.australianhumanitiesreview.org/archive/Issue-April-2006/EcoRigby.html. Accessed 15 February 2017.

Rigby, Kate 2008. 'Noah's Ark Revisited: (Counter-)Utopianism and (Eco-)Catastrophe'. In *Demanding the Impossible: Utopia and Dystopia*. Ed. Andrew Milner, Matthew Ryan and Simon Sellars. Special issue of *Arena Journal* 31. North Carlton: Arena Publications, 163–78.

Rigby, Kate 2009. 'Dancing with Disaster, *Australian Humanities Review* Issue 46 (May). www.australianhumanitiesreview.org/archive/Issue-May-2009/rigby.html. Accessed 15 February 2017.

Rigby, Kate 2014. 'Romanticism and Ecocriticism'. In *Oxford Handbook of Ecocriticism*. Ed. Greg Garrard. Oxford: Oxford University Press, 60–79.

Rigby, Kate 2015a. 'Earth's Poesy: Romantic Poetics, Natural Philosophy, and Biosemiotics'. In *De Gruyter Handbook of Ecocriticism and Cultural Ecology*. Ed. Hubert Zapf. Berlin: De Gruyter, 45–64.

Rigby, Kate 2015b. *Dancing with Disaster: Histories, Narratives, and Ethics for Perilous Times*. Charlottesville: University of Virginia Press.

Rose, Deborah Bird, with Sharon D'Amico et al. 2002. *Country of the Heart: An Indigenous Australian Homeland*. Canberra: Aboriginal Studies Press.

Seddon, Mark 2010. 'A Fresh Outbreak of Dutch Elm Disease is Threatening the Existence of the UK's Remaining English Elm Trees'. *BBC Newsnight*, 21 August. http://news.bbc.co.uk/2/hi/programmes/newsnight/8912727.stm. Accessed 15 February 2017.

Soper, Kate 2011. 'Passing Glories and Romantic Retrievals: Avant-Garde Nostalgia and Hedonist Renewal'. In *Ecocritical Theory: New European Approaches*. Ed. Axel Goodbody and Kate Rigby. Charlottesville: University of Virginia Press, 17–29.

Soper, Kate, Martin Ryle and Lyn Thomas (eds) 2009. *The Politics and Pleasures of Consuming Differently*. Houndmills: Palgrave Macmillan.

Steffen, Will 2004. *Global Change and the Earth System: A Planet under Pressure*, IGBP Book Series. Berlin: Springer-Verlag.

Sustainable Development Solutions Network Leadership Council 2013. 'An Action Agenda for Sustainable Development: Report for the UN Secretary-General', 6 June. http://unsdsn.org/resources/publications/an-action-agenda-for-sustainable-development/. Accessed 15 February 2017.

Uexküll, Jakob von 2010. *A Foray into the Worlds of Animals and Men, with a Theory of Meaning [1934]*. Trans. Joseph D. O'Neill, intro. Dorion Sagan, afterword Geoffrey Winthrop-Young. Minneapolis: University of Minnesota Press.

United Nations General Assembly 2005. '2005 World Summit Outcome: Resolution Adopted by the General Assembly', 24 October, A/Res/60/1. www.refworld.org/docid/44168a910.html. Accessed 15 February 2017.

Washington, Chris 2014. 'John Clare and Biopolitics', *European Romantic Review* 25 (6): 665–82.

World Commission on Environment and Development 1987. *Our Common Future* (Brundtland Report). Oxford: Oxford University Press.

4

Recycling materials, recycling lives: cardboard publishers in Latin America

Lucy Bell

Latin American *editoriales cartoneras* are small, independent publishers that make their books by hand out of recycled cardboard and aim to sell them at prices lower than those of large publishing houses. This cultural movement first began in Buenos Aires in the wake of the 2001 economic crisis, during which unemployment rates soared and people had a home one week but were homeless the next. One of the most visible impacts of the deep recession was the appearance of thousands of *cartoneros*, who took to the streets to collect scrap materials to sell on to individuals or recycling companies. In 2003, the founding members of Eloísa Cartonera began buying cardboard from the *cartoneros* at five times its market value to produce hand-painted books. Since then, the concept and practices of cardboard publishing have been adopted and adapted all over Latin America: in South America (Chile, Peru, Uruguay), but also as far north as Mexico, which now has the most *editoriales cartoneras* (approximately twenty active projects). Their shared identity as *cartoneras* conceals a wide range of different aims – literary, social, cultural, political and economic. A number of affinities, however, tend to unite them: the resistance to large commercial publishers that follow the logic of the capitalist, neoliberal economy; the desire to circulate the works of younger, lesser-known, more avant-garde authors; the need to reach a wider and different public in countries where books are very expensive and reading is often regarded as the activity of the 'privileged few'. Though many *cartonera* organisations collect the recycled materials themselves, others, like Eloísa, have strengthened their links with waste pickers or recycling cooperatives. In one case in São Paulo, which I shall explore in depth here, the publishing workshop and processes have been fully integrated into an existing recycling plant.

Why examine *editoriales cartoneras* in the context of a collection on literature and sustainability? Firstly, it is an intriguing instance of literary production in which the so-called 'three pillars' of sustainability – the environmental, the social and the economic – are invoked, intermeshed and transformed not (or not only) within the texts themselves, but most importantly in the processes of production. Secondly, it is a cultural movement in which the man-made material is a central player: cardboard is the element that unites these small publishers who work in different countries, contexts and conditions. In what follows, I shall examine the interplay between the human and the nonhuman, the social and the material, the economic and the environmental in the cardboard publishing movement. My contention is that, since the *cartonera* movement is one in which social, environmental and economic factors are intermeshed, it renders problematic the concept of sustainability as three distinct 'pillars'.

The 'three pillars' – suggesting solid, separate columns supporting a larger structure – were promoted at the Rio Earth Summit in 1992. The logic of the model can be seen in the structure of Agenda 21, the action plan that resulted from the summit (United Nations 1992), which is divided into two sections: the first, 'Social and Economic Dimensions', deals with issues such as reducing poverty, changing patterns of consumption and promoting health; the second, 'Conservation and Management of Resources for Development', focuses on environmental challenges like protecting the atmosphere, reducing deforestation, controlling pollution and managing radioactive waste. A decade later, the concept of the three pillars was adopted in the 2002 Johannesburg Plan of Implementation (JPOI), which aimed to 'promote the integration of the three components of sustainable development – economic development, social development and environmental protection – as interdependent and mutually reinforcing pillars' (United Nations 2002: 2). In spite of notions of interdependence and integration, the three components remain distinct. In fact, the UN continues to define sustainable development in terms of 'three pillars': 'economic development, social development and environmental protection' (United Nations 2010). This model has been recurrent in the discourse of, and scholarship on, sustainability over the last two decades, though there has been an increasing awareness of the need to integrate the three components.[1] Sometimes the 'three pillars' take on other guises, like the prominent idea of the 'triple bottom line'. This term was coined in 1994 by John Elkington, the founder of a British consultancy called SustainAbility, and taken up widely following the publication of *Cannibals with Forks: The Triple Bottom Line of 21st Century Business* in 1997. The 'triple bottom line' refers to a corporate model which demonstrates how businesses can – and must – thrive not only by generating profit but also by protecting the environment and fostering social justice.

Alternative models have of course been proposed. Some have suggested the need for a fourth dimension: in 1997, the German Wuppertal Institute reformulated the principles of sustainability as a prism in which the fourth triangle is the 'institution' (Dalal-Clayton and Sadler 2014: 54); in 2008, the Committee on Culture of the United Cities and Local Governments asked for culture to be recognised as the 'fourth pillar' of sustainable development', as drawn out in the Agenda 21 for culture (United Cities and Local Governments 2008). Others have sought to represent the model of sustainable development as an egg (see Kate Rigby's essay in this collection), as concentric circles, as interlocking circles, as a pyramid, and so on. One way or another, though, these models are all based on the possibility of differentiating between human and nonhuman, cultural and natural.

This possibility has been powerfully challenged in an emerging branch of the environmental humanities known as 'new materialism' (a concept coined by Manuel DeLanda and Rosi Braidotti in the 1990s). It is headed by scholars like Stacy Alaimo, Karen Barad, Jane Bennett and Nancy Tuana, and although their recent works diverge in many ways, they all call for a more nuanced understanding of the relationship between the human and nonhuman, and for an acknowledgement of the agency of matter – that is, physical substance, whether 'naturally' occurring or man-made. Jane Bennett's theory in *Vibrant Matter* – which derives from a long lineage of materialist philosophers from Democritus and Epicurus, through Spinoza and Thoreau, to Deleuze and Guattari – is that nonhuman things can 'act as quasi agents or forces with trajectories, propensities, or tendencies of their own' (2010: viii), and that they constitute actants in the web of forces that produce political situations or problems. Tuana views Hurricane Katrina as a call for the urgent need to acknowledge the viscous porosity of the human and nonhuman world, the natural and the social ('viscous' because these elements are neither solid nor fluid, but rather something in between) (2008: 193–4). She argues the necessity of an 'interactionist ontology' to understand a phenomenon that resulted from a complex interaction between 'natural' and 'man-made' factors – low pressure and warm ocean temperatures, but also deforestation, industrialisation and climate change – and whose material impacts are inseparable from social issues of class, race and poverty. Alaimo, for her part, proposes the notion of 'trans-corporeality' as a theoretical model that allows us to rethink agency beyond the sphere of the human, to understand 'the often unpredictable and unwanted actions of human bodies, nonhuman creatures, ecological systems, chemical agents, and other actors' (2010: 2). Trans-corporeality is defined simply as the movement across bodies but challenges us to delve into the complex and often disturbing material interconnections between the human body and the more-than-human world. Her thinking, like that of Bennett and Tuana, undermines the separability of the 'three pillars of sustainability' since it draws attention to the 'often invisible, but nonetheless material, flows of substances and forces between people, places, and economic/political systems' (Alaimo 2010: 9).

In a more recent article, Alaimo (2012) engages with the discourse of sustainability from a new materialist standpoint. She critiques the standard definition of sustainable development from the 1987 Brundtland Report: development that 'meets the needs of the present without compromising the ability of future generations to meet their own needs' (World Commission 1987). For Alaimo, this oft-quoted definition reveals an overarching anthropocentric logic, since the 'generations' are normally interpreted as human, and the nonhuman world is reduced to a resource for meeting human needs (2012: 562). The prevailing thinking on sustainability, from this perspective, is counter-productive insofar as it epitomises a technocratic, scientific epistemology that separates human selves, needs and practices from nonhuman environments, resources, and problems. Accordingly, she suggests that the role of the environmental humanities is to 'formulate more complex epistemological, ontological, ethical, and political perspectives in which the human can no longer retreat into separation and denial or proceed as if it were possible to secure an inert, discrete, externalised this or that' (Alaimo 2012: 563). From this perspective, the discourse of sustainability can only be productive if it is built not upon binary thinking that opposes nonhuman environment with human society, but rather from a new materialist perspective in which the trans-corporal flows between the human body and the environment are acknowledged (Alaimo 2010, 2012), matter is appreciated in all its vibrancy and agency (Bennett 2010) and phenomena come into being through the interaction between social, biological, natural and human-made components (Tuana 2008).

In what follows, the Latin American cardboard publishing movement will be read as the result of a complex interaction between social, material, environmental and economic components. This will be illustrated through an in-depth analysis of a publication by the São Paulo-based publishing cooperative Dulcinéia Catadora entitled *Catador (Waste Picker)*, a collage of texts about the recycling cooperative Cooperglicério, of which Dulcinéia is part, and where its workshop is located. As we shall see, the physical substances with which the waste pickers work in this collection and the practices it represents – the waste collected from streets and rubbish tips, the air and noise pollution by which they are surrounded, and of course the cardboard that they collect – are nonhuman materials that are humanly created. From the outset, then, the practices illustrated in this collection problematise the separation between material and social, environmental and human in a way that both resonates with, and challenges, new materialist thought. Whereas Alaimo, Bennett and Tuana blur these boundaries in order to raise the standing of the more-than-human world, the *catadores* use this liminal zone as a site of social denunciation and resistance: they highlight and critique their connection with the waste materials they collect, and seek to (re)generate their collective identity through the regeneration of materials.

The connection between waste and waste picker will be brought into dialogue not only with new materialist thought, but also with Zygmunt Bauman's theory of 'wasted lives' (2004). For Bauman, the supposedly forward-moving trajectory of modernity – and the relentless forces of order-building and economic growth it entails – has brought about a division of society into producers and consumers on the one hand, and outcasts (job-seekers, asylum-seekers, beggars, migrants, scavengers, etc.) on the other. The latter lack the skills, means or opportunity to participate in labour and consumer markets, and therefore become marginalised from society, inheriting the properties of waste materials, of things deemed 'useless' and disposable (Bauman 2004: 12–13). In the English language, 'to be "redundant" means to be supernumerary, unneeded, of no use' (Bauman 2004: 12). 'Redundancy', Bauman points out, 'shares its semantic space with "rejects", "wastrels", "garbage", "refuse" – with *waste*' and 'the destination of waste is the waste-yard, the rubbish heap' (2004: 12; author's italics). This division between valuable and worthless, included and excluded, is not a natural one, but rather a barrier erected by humans as a strategy to ensure the smooth running of the status quo.

Bauman's theory is clearly grounded in the discipline of sociology and has as its principal concern the connection between economic progress and social inequalities. In the introduction to his book, he makes his disciplinary position clear by explaining that his point that 'the planet is full' is 'not a statement in physical or even human geography', but 'in sociology and political science' (Bauman 2004: 5). However, it is noteworthy that the notion of 'wasted [human] lives' is derived from that of material waste. Indeed, he opens his sociological study by presenting the human inclination to produce waste – which correlates with the heightened consumerism that characterises modernity – through the image of Leonia from Italo Calvino's 1972 novel *Invisible Cities*. Leonia's inhabitants have a thirst for novelty that creates 'a fortress of indestructible leftovers', which dominate the city 'on every side, like a chain of mountains' (cited in Bauman 2004: 2). And this image functions as a powerful illustration of the elaborate theory of 'human waste' or 'wasted humans' developed throughout Bauman's work. To a certain extent then, we shall see that Bauman's theory resonates with new materialist thought insofar as it shows the connection between human and nonhuman waste in the workings of modernity.

Working with waste, either delving into public bins or diving into municipal dumps, the urban scavenger might be seen to epitomise the 'wasted life', associated as he or she is with the waste he or she collects, and suffering from the same rejection, exclusion and invisibility. However, my contention is that the strategy of exclusion condemned by Bauman is transformed into strategies of inclusion by the Coopergliçério waste pickers and publishers – in their activities, their productions and in the book itself, which has a performative function. By adding value to waste materials, they not only valorise recycling processes but also their own activities, livelihoods and lives.

A complex phenomenon: environmental, social and economic factors

Since it encompasses a vast geographical area and a wide range of motivations, it can be difficult to understand what ties the *cartonera* publishing movement together. As suggested by the names of the publishers – Eloísa Cartonera, La Sofía Cartonera (Argentina), Sarita Cartonera (Peru), La Verdura Cartonera, La Cartonera Cuernavaca (Mexico), and so on – the most obvious shared element is the material used to make the books: recycled corrugated cardboard (*cartón* in Spanish), used because it is widely available, and also relatively malleable, solid, durable and lightweight. Yet the adjective '*cartonera*' is ambiguous, since it refers not only to the material *cartón*, but also to the *cartonero* figure, the cardboard collector or waste picker who played a central role in the formation of the first cardboard publishing organisation, Eloísa Cartonera, and continues to work directly with some publishing organisations (like Eloísa and Dulcinéia Catadora) to this day. The term *cartonera* therefore binds together the human and nonhuman, encompassing the complexity of a publishing movement connected not only with cardboard materials and *cartonero* workers, but also with broader socio-economic contexts of unemployment, poverty and social exclusion. This semantic ambiguity is attested by the different names taken on by cardboard publishing projects in Portuguese-speaking countries: whereas in Mozambique Kutsemba Cartão (Kutsemba Cardboard) identifies with the cardboard material itself, in Brazil, Dulcinéia Catadora and Katarina Katadora identify themselves with the figure of the waste picker (*catador* in Brazilian Portuguese), reflecting the fact that recycling, particularly in the developing world, is not only an environmental issue, but also an economic and social one.

Waste disposal and recycling clearly pose a huge environmental challenge on a global scale, but some of the problems are magnified in an area now referred to as the 'global South'. According to Martin Medina (2005), although a large chunk of municipal revenues are spent on municipal solid waste, most Latin American municipalities collect only a fraction of it – in many cases between 50 and 80 per cent. Preoccupied as their local governments are with extending waste collection, most Latin American cities do not have formal recycling programmes. In Brazil, for example, over half of all solid disposable waste is inadequately disposed of in open-air dumps and only 32 per cent of municipalities have any form of segregated waste collection scheme (IBGE 2008). And, as Medina explains, methods of disposal and recycling that are appropriate in the developed world can be completely inappropriate when applied to cities in the developing world.

This environmental situation, combined with high levels of poverty and an abundance of unskilled workers, has led to a wide proliferation of informal recycling activities in Latin America and the Caribbean which, according to recent estimates, are a means of survival for between 400,000 and 4 million people (IADB 2014). These informal waste pickers are to thank for a large proportion of the recycling activities undertaken in Latin America. In Brazil, for example, though only about one third of municipalities have any form of waste separation scheme, the majority of certain materials get recycled: between 70 and 80 per cent of cardboard and 98.3 per cent of aluminium cans (making it the world leader for the recycling of this particular material) (PNSB 2008). This disparity is due to the fact that the majority of recycling activities are carried out informally by waste pickers, either working individually or in cooperatives, for a small payment.

The social aspect of the cardboard publishing movement has gone far beyond the context of waste picking, however. A key impetus behind the movement is to make books accessible to more people – to *create* new reading publics. In most Latin American countries, paperbacks are very expensive and minimum wages are very low, and the price of a book can represent up to a fifth of a monthly wage. This is one of the reasons behind the lack of a well-established reading culture in these countries in spite of relatively high levels of literacy. In Mexico, for example, a 2012 survey found that 54 per cent of Mexicans do not read – even though only about 7 per cent of the country is illiterate (Fundación Mexicana 2012). *Editoriales cartoneras* sell more affordable books by reducing the costs of production in different ways: they use cheap, recycled materials; their workforce, apart from the cardboard collectors themselves, is largely composed of volunteers; and authors donate their works to them free of copyright costs. The cost reduction is a significant gesture towards dissolving the barriers that render literature exclusive and elitist. The motivation for recycling materials, therefore, cannot be separated from the desire to open literary production to a wider reading public.

It should also be pointed out that the process of upcycling (increasing the value and quality of the original object) is crucial to the economic self-sustainability of *cartonera* projects. For all *cartoneras*, financial autonomy is paramount. Aurelio Meza, for example, affirms that the organisation he has been involved in, Kodama Cartonera (Tijuana), is financed by funds from the participants' teaching and research. This is because they believe that 'state financing for an independent project means that it gradually loses its independent quality (economically, creatively and ideologically)' (Meza 2014). In the same vein, Eloísa runs off the proceeds of its sales, and on several occasions its members have turned down governmental financial backing for micro-businesses in order to retain financial independence. Turning a piece of discarded cardboard into an artistic object and turning a piece of writing (which may have been rejected or be deemed 'unfit for publication' by a large publishing house) into a publication increases the value of the original materials, generates a profit and helps publishers maintain their financial autonomy.

The cardboard material and the material process of recycling therefore constitute a site of intersection between cultural, economic and social processes. This can be seen in the self-description by Dulcinéia:

> Always in contact with Movimento Nacional dos Catadores de Reciclagem (National Movement of Cardboard Pickers), the group values the work of cardboard collectors, acts in defence of social inclusion and intends to develop the creative and artistic potential of its members. As a daily praxis, the group buys cardboard boxes collected in the streets by street pickers called 'carroceiros' at a price higher than that paid at 'sucatas' (commercial outlets that buy scrap). These boxes are cut and then the pieces of cardboard are painted one by one with gouache. Each book is sold for 6 reais and income is shared among participants. It is a model of creative economy, to publicize the work of new and unknown writers is also one of the main goals of Dulcinéia Catadora. (Dulcinéia Catadora n.d.)

Value, here, has multiple simultaneous facets: environmental (referring to the importance of recycling in the context of dwindling resources and problems of litter, landfill sites and contamination); social (relating to the worth and position of waste pickers in Brazilian society); and economic (indicating the financial value of the collected materials and end-products). For this organisation, the process of recovering waste materials from streets, tips and bins is inseparable from the activity of promoting social inclusion and artistic creativity. Transforming a discarded object into a useful, valuable object involves dismantling the barriers between inside and outside that are erected by humans to create social and cultural distinctions and exclusions. This transformation, therefore, also entails bringing into circulation writers who have not yet been published and selling the books to those who cannot normally afford them.

A case study: socio-material composites in *Catador*

In order to dig more deeply into socio-material interactions in the *cartonera* movement, I will now turn to a text directly concerned with these issues: *Catador* (Dulcinéia Catadora 2012). The environmental and human aspects of recycling are absolutely central to this project: it was originally set up in 2007 by the artist and translator, Lúcia Rosa, and the son of a waste picker, Peterson Emboava; it is run principally by waste pickers from the recycling cooperative Cooperglicério and their relatives; and the books are now produced in a small office on that recycling site. The book consists of a selection of interviews with members of Cooperglicério interspersed with other texts by local collaborators (a rap artist, a filmmaker, a graffiti artist, the development manager of a local public health institute and an academic from the University of São Paulo), all reflecting on the work, lives and desires of the *catador*. It was produced collectively by the members of the Dulcinéia cooperative – Ana d'Angelo, Andréia Ribeiro Emboava, Eminéia Silva Santos, Elizângela Juventino, Lúcia Rosa and Maria Dias.

In their introduction, the writers present the collection as a 'book of tales, memoirs and images of the Glicério Cooperative' (Dulcinéia Catadora 2012: 7).[2] The story of the cooperative is told in a series of first-person responses to interview questions. Structurally, the book is not separated into the different lives of the waste pickers, but rather into the different questions asked by the coordinators. In that sense, it might be viewed not as a collection of interviews, but rather as a collective autobiography of the cooperative, thus opening up the genre of biography beyond the writing of an individual's life to that of a whole community. Moreover, this collective encompasses not only human lives, but also nonhuman materials and spaces. The Glicério Cooperative refers both to the organisation run by the *catadores* and to the site on which the waste is sorted: the recycling plant underneath a highway flyover in the Liberdade area of São Paulo. The environment in which, and with which, the waste pickers work is therefore inseparable from their lives and livelihoods.

Hence the photographs in the collection (the 'images of the Glicério Cooperative') are of the Dulcinéia waste-pickers-publishers-interviewers-writers, of the site, and of the collected waste materials (at various stages of the sorting process). The publishers are pictured on the Glicério site, Eminéia with her baby, Andréia, painting a cardboard cover with waste in the background and Maria in front of bin bags and waste trucks (Figure 4.1).

The flattening effect of the photograph highlights the interconnection between human bodies and physical environments. In terms of graphic design, an image of torn cardboard accompanies each section, either as a margin or as the background, which constantly reminds the reader of the cardboard cover of the book they are reading – and holding (Figure 4.2). As we shall see, the texts in this collection, like the images, centre around the relationships between humans (those who produce waste and those who collect it), nonhuman materials (the waste itself or the recycled object) and spaces (the road along which the waste pickers' cart is pulled or the recycling cooperative in which the rubbish is gathered). Albeit indirectly, the supposed 'three pillars' of sustainability are invoked in a way that makes it very difficult to uncouple environmental issues of waste and recycling from social problems and economic processes.

Figure 4.1 Photographs of the publishers: *(left to right, top to bottom)* Maria Dias, Andréia Ribeiro, Elizângela Juventino, Eminéia Silva Santos (Dulcinéia Catadora 2012: 7)

The social issues associated with the activity of waste picking are a recurrent theme in *Catador*. The introduction by Maria, Andréia, Elizângela and Eminéia underlines this: 'People show indifference and even a lack of respect when they see a cart in the middle of the road' (7). This cart is a synecdoche for the rubbish that is piled into it and for the waste picker who pulls it; the disregard and disrespect with which it is treated is thus

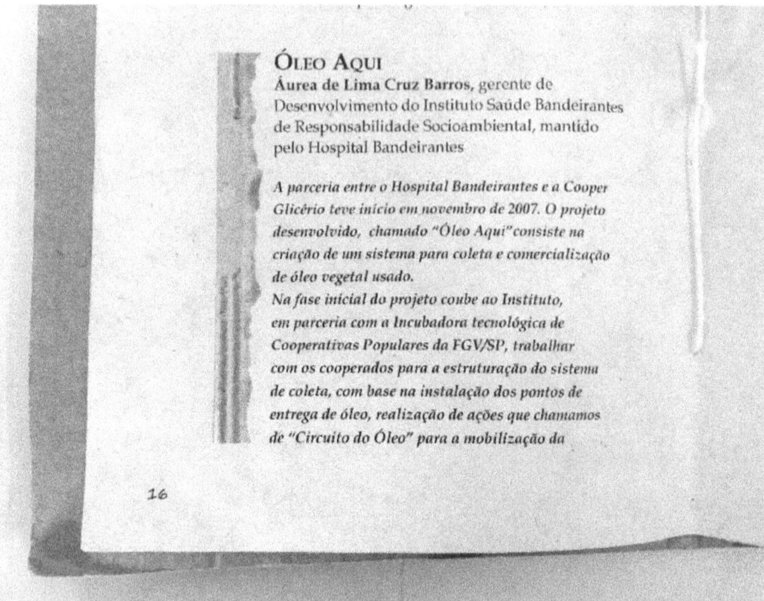

Óleo Aqui

Áurea de Lima Cruz Barros, gerente de Desenvolvimento do Instituto Saúde Bandeirantes de Responsabilidade Socioambiental, mantido pelo Hospital Bandeirantes

A parceria entre o Hospital Bandeirantes e a Cooper Glicério teve início em novembro de 2007. O projeto desenvolvido, chamado "Óleo Aqui" consiste na criação de um sistema para coleta e comercialização de óleo vegetal usado.

Na fase inicial do projeto coube ao Instituto, em parceria com a Incubadora tecnológica de Cooperativas Populares da FGV/SP, trabalhar com os cooperados para a estruturação do sistema de coleta, com base na instalação dos pontos de entrega de óleo, realização de ações que chamamos de "Circuito do Óleo" para a mobilização da

16

Figure 4.2 Page design of *Catador*, including photograph of torn cardboard as margin

simultaneously for the waste picker and for the waste itself. The theme of marginalisation is explored further in the rap song 'Cata as dores de papel' by local rap artist Fábio Prestes – a song that effectively becomes a poem within the written body of the text, though the reader is referred to the song through a YouTube URL (www.youtube.com/watch?v=T8IlAfwmMzE). A play on words meaning 'Waste paper pickers', 'Paper waste pickers' and 'Collect paper pains', the title cleverly combines the different intermeshed concerns of the song, notably the merging of human subject and waste object, and the physical suffering of the *catador*.

Indeed, the process of *becoming waste* is enacted in the song's movement from simile (and assimilation) to embodiment. In the first half, the first-person voice affirms: 'I'm in the streets crushed between cars / I am like [or I look like] the cardboard that was thrown into my cart' (22). *Pareçer* means to resemble both in appearance (to look like) and in nature (to be similar to, to be like). In the first sense, the waste picker's physical appearance is compared to that of cardboard, suggesting that he is thin, due to poverty and hunger, and dirty, through exposure to car fumes and contact with unseparated rubbish. In the second sense, the verb turns the cardboard into a simile for the figure of the waste picker, whose whole being becomes connected with the waste materials which he collects – though the simile still allows some distinction between the discarded cardboard material and the waste picker who salvages it. Later, the titular wordplay reappears in the body of the song: 'Cata as dores de papel', that is, 'collectors of paper' and 'collectors made of paper'. The semantic undecidability here fuses the two meanings: the human collecting waste and the human who embodies or incorporates it.[3]

This fusion can be interpreted in the context of Bauman's 'wasted lives'. For Bauman (2004: 12–13), anyone who is outside the society of producers and consumers suffers from a dual threat, biological and social survival. The most obvious example is the unemployed, a person who in the English language is 'redundant'. Bauman quotes Danièle Linhart, who in *Loss of Job, Loss of Self* suggests that the unemployed find themselves not only devoid of an income, but also 'stripped of their dignity as workers, of self-esteem, of the feeling of being useful and having a social place of their own' (Bauman 2004: 13). They become social rejects, destined for the rubbish heap of wasted lives. As shown in *Catador*, the scavengers' material connection with waste – their bodily contact with it and their intrusive, excessive presence on the highway – corresponds with a social connection, as they are cast aside, alienated, and looked down upon:

> Gated residences, barbed wire,
> security cameras all around
> Today I saw, today I saw, today I saw you
> in the TV screen's reflection, sitting in the guard house
> ... your limit is the corner and mine is far away. (22)

Social barriers are materialised by man-made physical barriers here: the waste pickers are separated from 'worthy' or 'valued' citizens by gates, barbed wire, walls, windows, limits and corners, but also by the lens of the security camera and the television screen. The contact between the inhabitants of the gated residences (who discard their waste) and the *catadores* (who pick up their waste) is reduced to the reflection in a television screen. Implicit in the security camera and the guard house is not only the exclusion, but also the criminalisation of the *catador*, and thus the barbed wire and gates suggest those surrounding a prison.

On one level, these connections are semantic: while Bauman's theory is underpinned by the etymological and semantic connections between redundancy (unemployment) and garbage, Prestes's song/poem is based on wordplay. On another level, these semantic connections reproduce deep ontological connections: the process of *becoming waste* that occurs throughout the song/poem might be regarded as an enactment of what Tuana terms an 'interactionist ontology' (2008: 188), meaning that beings are not self-enclosed entities, but rather ever-unfolding products of their interactions with other beings and the more-than-human world. Humans *become waste* through complex interactions with nonhuman phenomena, through the merging of human identity and nonhuman waste, social rejection and physical exclusion, economic forces and material factors. In Prestes's song/poem, what Tuana terms the 'materiality of the social' (2008: 210) is inseparable from the sociality of the material, since social categories become fused with material categories and social barriers become inextricable from physical barriers.

As suggested by the title, physical suffering is another recurring theme in Prestes's song, which depicts conditions of poverty, difficult working conditions and long hours. These social issues are presented by Prestes in all their physical manifestations of hunger, exhaustion and pain:

on the walk of fame hunger goes on
In an enormous present in a wasteland
if it's hot your feet burn they burn if it's cold too
they're families on the street without notion of a schedule. (22–3)

Time and space become vast, empty territories marked only by the pangs of hunger and the pain of burning feet. The song's unrelenting pace reproduces that which in Alaimo's terms might be referred to as the trans-corporeal relationship between the external world of the city and the internal world of the body: the road ironically referred to as the 'walk of fame' fuses with the raw pain of the burning feet; the urban wasteland (suggesting an isolated rubbish dump or empty, early morning roads) merges with the empty stomach and the unending hunger (the internal body clock that substitutes any time schedule). Furthermore, the waste picker's extreme hunger (which makes him or her 'paper thin,' as suggested above) hints at the fact that he or she might be 'wasting away,' starving, and ultimately dying, thus evoking another way in which humans merge with the nonhuman – through the organic process of becoming waste. Yet whereas Alaimo's notion of trans-corporeality 'denies the subject the sovereign, central position' (2010: 16), Prestes's song is focused on human suffering, on the constant, cumulative, invisible pain of the burning feet through the contact with tarmac, concrete or earth, from pushing a cart on foot while others drive past in cars and on motorbikes. This pain cannot be tied down to a single time and location, but rather expands across time and space, hence the 'enormous present' and the loss of any 'notion of a schedule.'

Another aspect of the suffering undergone by the waste picker in the text is that which results from his contact with noise and air pollution: 'I survive another day in the midst of the pollution, / directing the orchestra of beeping horns ...' (22). As suggested by the verb 'survive,' the life of *catadores* is precarious, not only because they have to dart between fast-moving vehicles, but also because of the fumes that penetrate their bodies – skin, eyes, nose, throat, lungs – causing burning, irritation and disease. Prestes's depiction of the waste picker underlines the 'substantial interconnections between humans and the wider world', the 'movement across human corporeality and nonhuman nature' (Alaimo 2010: 3). As Alaimo puts it, 'the human body is never a rigidly enclosed, protected entity, but is vulnerable to the substances and flows of its environments' (2010: 28). Moreover, at the forefront of this rap song is the matter of environmental injustice and inequality, the idea that the social exclusion to which the waste pickers are subjected also emanates in material disadvantage. This hints at a broader social situation raised by Prestes: that those who deplete the most resources, represented here by the inhabitants of the gated communities, are also (but only partially and superficially) protected from the ugly and potentially harmful waste produced by their consumption. Environmental damage, perversely, impacts disproportionately on the poor – those people excluded from the society of producers and consumers – who are least responsible for producing it.

'Cata as dores', however, is not just a passive reflection of socio-environmental injustice, but also a performative act of resistance. In its written version, two phrases stand out because they interrupt the flow of the poem through exclamatory caesurae: 'But pay attention!' and 'So stop! Breathe in the fumes' (22). The reader/listener is called upon to engage with the issues at stake not just cognitively (by 'paying attention' and thinking, by 'stopping' to reflect), but also experientially and sensually: to breathe in the fumes invokes the sense of smell, as well as that of taste; to 'pay attention' is both to see (the words on the page, but also the waste and *catador* they depict) and to listen (to the song and to the racket of the city's traffic and car horns it describes). The reader, rather than remaining detached, is physically entangled in the poetic matter and material of the song.

This sensorial invocation is reproduced in the cardboard cover of the book: the physical connection between reader, cardboard and waste picker through the senses of sight, touch and smell brings the social barriers depicted in the song/poem crashing down. The cover, like the words inside, is a call for the reader to stop, to observe and to pay attention not only to the material object itself, but also to the context in which it was produced. It therefore supports the performative dimension of the book, its function as an advertisement for the cooperative and a call for various forms of citizen and governmental participation: for donations of recyclable materials (8); for locals to separate their waste into wet and dry materials, and to bring recyclables directly to the cooperative (24); and for the local government to recognise and support their work (27–8).[4] The trans-corporeal flows between human body and nonhuman matter are therefore used as a tool by the *catador*-publishers to invoke the participation of the reader, as a call for action and collaboration.

(Un)equally environmental perspectives: material agency and human activism

If the new materialists' project is to draw attention to the vitality and quasi-agency of matter, to the ability of the more-than-human environment to produce social, political and economic effects, the aim of the *catador*-publishers of Cooperglicério is to affirm their social participation, to strengthen their collective identity and to recycle their 'wasted lives'. This brings us to an important problem, namely that we might be faced with an irreconcilable difference between full and empty belly sustainabilities, which correspond with Ramachandra Guha's and Joan Martinez-Alier's 'full-stomach' and 'empty-belly' environmentalisms (1997: xxi). In order to address this problem, it will be useful to bring into dialogue Bennett's reflections on waste and Alaimo's take on sustainability with one of the texts from *Catador* which describes a 'sustainable' project that aims both to recycle discarded (and potentially noxious) oil and to raise money.

Waste is used by Bennett as her first case study of 'thing-power'. She begins by describing her personal experience of encountering a pile of litter on the floor one morning in Baltimore:

> as I encountered these items, they shimmied back and forth between debris and thing – between, on the one hand, stuff to ignore, except insofar as it betokened human activity (the workman's efforts, the litterer's toss, the rat-poisoner's success), and, on the other hand, stuff that commanded attention in its own right, as existents in excess of their association with human meanings, habits, or projects. (Bennett 2010: 4)

She depicts the *effect* of the rubbish on her, as she is repelled both physically by the dead rat and morally by the human action of littering. But she also becomes aware of the singularity and force of these waste objects, of their 'ability to make things happen, to produce effects' independently of human intervention (Bennett 2010: 5). This vibrancy is then illustrated by a passage by Robert Sullivan, who portrays the garbage hills outside New York as live and lively assemblages: 'the garbage hills are alive … there are billions of microscopic organisms thriving underground in dark, oxygen-free communities', cells that 'exhale huge underground plumes of carbon dioxide and of warm moist methane' (cited in Bennett 2010: 6). Bennett no doubt chooses rubbish as her first illustration of thing-power as it is the ultimate example of vital materialism, for *even* the discarded object – the object no longer seen as *useful*, and therefore at the bottom of the hierarchy of value – in fact has power and agency.

The active power of waste materials is acknowledged in *Catador* in a contribution about a project called 'Oil Here', written by Áurea de Lima Cruz Barros, one of the organisers of the project and development manager at the Bandeirantes Health Institute in São Paulo. Launched in 2007 to set up a system for the collection of used vegetable oil for commercial resale, the project was a collaborative effort by Cooperglicério and the Health Institute. Cruz Barros presents discarded cooking oil as a 'polluting agent' that can be seriously harmful to the local environment: 'research shows that one litre of oil can pollute thousands of litres of water' (17). The oil, once discarded, does not simply cease to exist, but rather continues to move, to invade other substances and to penetrate human bodies through water pollution.

However, Cruz Barros's emphasis – unlike that of Bennett – is on the fact that *material* agency can productively be utilised by a corresponding *human* agency. As she puts it, one of the aims of the project is to 'contribut[e] to local sustainable development thanks to the creation of a mechanism that transforms a polluting agent into a financial resource ... It is possible for us to turn environmental passives [*passivos ambientais*] into local actives/resources [*ativos/recursos locais*]' (17). The oil – the 'polluting agent' – is valued in human terms, in the fact that it can be used as a 'financial resource', a good that can be sold for profit, to generate wealth. Cruz Barros encourages the reader or citizen to engage in a shift in thinking, allowing for so-called 'environmental passives' – waste objects that have been excluded from use and become devoid of purpose – to be transformed, through human intervention, into valuable materials. Again, this highlights the priority given to *human* needs, agency and power in this project.

Another stated goal of the project is the 'development of the role of the waste picker as environmental agents and people who bring a solution to the collection of this pollutant' (17). Here, it is not the waste material but the waste picker who becomes the 'environmental agent'. The ecological problem of waste and pollution, then, becomes an economic solution for the *catador*. Furthermore, the revaluing of waste is not only material and economic, but also social, since the 'Oil Here' project promotes a shift from waste picker to 'environmental agent'. The collaboration thus helps fight against social exclusion, against 'wasted lives', by transforming (passive) rejects from the socio-economic system into (active) participants, or agents. Moreover, Cruz Barros emphasises the importance of the autonomy of these environmental agents: in charge of the whole process from contacting donators to reselling the product, they receive all the income from the enterprise, hence the 'self-sustainability of the project' (17). Sustainability, in this context, is cast in very human terms, in terms of economic self-sufficiency, and human effort and reward (Figure 4.3).

Catador therefore foregrounds a division between 'rich' and 'poor' belly sustainability and offers a necessary corrective highlighting the social dimensions so often neglected in environmental discourse. Specifically, it forces us to ask some pressing questions about who the discourse and practices of sustainability belong to, and who its key actors are. In the

Figure 4.3 The waste pickers' cart transformed by Dulcinéia Catadora into a stall for their beautiful cardboard books at the third Feira Plana (São Paulo, 2014)

Brundtland Report's section on poverty, the question of *whose* sustainability is at stake is raised by Aristides Katoppo:

> I think this Commission should give attention on how to look into the question of more participation for those people who are the object of development. Their basic needs include the right to preserve their

> cultural identity, and their right not to be alienated from their own society, and their own community. So the point I want to make is that we cannot discuss environment or development without discussing political development. (World Commission 1987: 27)

Two key points become apparent here. Firstly, Katoppo underlines the fact that the concern with the 'environment-in-itself', the material world in its own right, cannot take place independently of an attention to basic human needs. As Bispo points out in *Catador*, these needs precede any discourse of sustainability: 'before they started talking about sustainability there were already *catadores* here. It was a group of people who lived on the streets who got together, collected the material and had to sell it to make money to buy food, to allow their families to survive' (14). At the heart of the Coopergliccério's concept of sustainability is survival – their own, but also that of their families and community. They collect waste, as they put it, in order to 'sustain/support our families [*sustentar nossas familias*]' (7). This is echoed later by Arivaldo, who insists that 'there will always be someone working in recycling to get the benefits, to sustain/support [*sustentar*] themselves' (13). Sustainability for the *catadores* means 'to be able to carry on providing sustenance', meeting their own basic needs and those of their children. More than an environmental challenge, it is a matter of human instinct, biology and finances. And, as we have seen, the *catadores*, like Bauman and Katoppo, view survival as social as well as biological and economic: from the perspective of the poor, the problems of nutrition cannot be separated from those of social inclusion and collective identity; the interconnection of social and economic with ecological sustainability is (for all its problems) necessary. Part of the discourse of sustainability is and *should be* about human sustenance.

Secondly, Katoppo brings to the fore the question of participation. As we have seen, the collaboration presented and reproduced in *Catador* proposes an alternative view to that which casts certain citizens as 'objects' of sustainable development, and others as 'subjects': what is emphasised and enacted in this collection is the coming-into-being of the *catador* as agent of sustainable socio-environmental development. This empowerment is a form of resistance to the phenomenon of wasted lives as perceived by Hauke Brunkhorst: 'for those who fall outside the functional system [of employment], be it in India, Brazil or Africa … all others soon become inaccessible. Their voice will no longer be heard, often they are literally struck dumb' (2001: 233). One of the aims of the Coopergliccério and their collective autobiography is to give the *catadores* a voice, to show that they each have different stories and singular experiences, whilst helping to solidify a collective identity. The material agency illustrated by 'Oil Here' therefore becomes a springboard for human activism.

This brings us to Alaimo's question: 'is it possible to recast sustainability in such a way that it ceases to epitomise distancing epistemologies that render the world as resource for human use?' (2012: 263). It is difficult to get around this problem without facing a division between a 'sustainability of the poor' and a 'sustainability of the rich.' In a sense, the *Catador* collection does recast the sustainability discourse insofar as it implicitly dismantles its 'three pillars,' undermining the clear differentiations between nonhuman and human upon which it has been built. From the perspective of the waste picker, material agencies are inextricable from human agencies, since recycling waste materials and urban wastelands simultaneously regenerates wasted lives. However, as Katoppo insists and the *catadores* in this collection imply, it is impossible to avoid seeing the world as a resource if one's daily struggle is that of biological, economic and social survival.

The subject (in both senses of the word) of sustainability is raised repeatedly in *Catador*. It is revealing that Bispo uses the impersonal form '*se*' ('Before they started talking about sustainability …') (14), thereby casting sustainability as a discourse, a theory, an institutionalisation that emerged much later than the practices that it theorises, which stem from necessity: the waste pickers *had* to sell recyclable materials to make money, to buy food, to live. Another contribution in *Catador*, by filmmaker Evaldo Mocarzel, highlights the fact that 'public-sector' sustainability has more recently been supplemented by 'private-sector' sustainability. He describes his documentary *In the Margins of Waste* (2011) as an attempt to 'dignify the crucial work of the *catador* at a historical moment in which environmental awareness has become big business' (Dulcinéia Catadora 2012: 10). This points to the fact that global brands like Coca-Cola, Walmart, Nestlé, McDonald's and Nike 'are defining sustainability and implementing it through their operations and supply chains to gain competitive advantages and increase sales and profits' (Dauvergne and Lister 2013: 1). The point of the documentary (and the *Catador* collection in which it is referenced) is to take ownership of sustainable processes, to take the discourses from the clutches of big business and recast them from the perspective of a small cooperative.

Catador therefore sheds light on the different components at work in so-called 'sustainable processes,' from the biological experience of hunger to commercial strategies of marketing. It simultaneously presents and produces some of the agencies combined in the activities of discarding, collecting and transforming waste, wherein biological and natural agencies are compounded with social activism, economic activities and cultural acts. In its integration of social, artistic, economic and material processes, this collection is emblematic of the broader *cartonera* movement, which constitutes a celebration of the cardboard material with which the books are made, upon which a collective identity is built, thanks to which a readership is created, and through which lives are recycled. Yet unlike in Bennett's account of thing-power, where discarded objects are viewed as things-in-themselves in all their singular vitality, the cardboard used by *editoriales cartoneras* is valued insofar as it serves a purpose for cardboard pickers and publishers, insofar as it is collected, cleaned, cut, decorated, stitched, bound, turned into a book, resold, 'read', passed on and shared. The colourful cardboard covers, containers of literary materials, are therefore a celebration of human creativity, of the ways in which different resources can be used to produce new forms – forms of art, resistance and empowerment.

Notes

1 Recent uses of the 'three pillars' include cases in the context of construction and engineering (Beheiry, Chong and Haas 2006); life cycle assessment (Swarr et al. 2011); sustainable development (Hansmann et al. 2012); and many more.
2 All subsequent references to *Catador* will be from this edition. All translations are my own.
3 This is very similar to the process incorporated in the 2010 film *Waste Land*, directed by Lucy Walker and starring the Brazilian artist Vik Muniz, though space does not allow me to elaborate on this connection here.
4 There does appear to be some tension – within this and other cardboard publishing organisations – between the desire to remain financially autonomous and the desire to gain governmental backing.

References

Alaimo, Stacey 2010. *Bodily Natures: Science, Environment, and the Material Self.* Bloomington: Indiana University Press.

Alaimo, Stacey 2012. 'Sustainable This, Sustainable That: New Materialisms, Posthumanism, and Unknown Futures', *PMLA* 127 (3): 558–64.

Bauman, Zygmunt 2004. *Wasted Lives. Modernity and its Outcasts.* Cambridge; Malden, MA: Polity Press.

Beheiry, Salwa M., Wai Kiong Chong and Carl T. Haas 2006. 'Examining the Business Impact of Owner Commitment to Sustainability', *Journal of Construction, Engineering and Management*, 132 (4): 384–92.

Bennett, Jane 2010. *Vibrant Matter: A Political Ecology of Things.* Durham, NC; London: Duke University Press.

Brunkhorst, Hauke 2001. 'Global Society as the Crisis of Democracy'. In *The Transformation of Modernity: Aspects of the Past, Present and Future of an Era.* Ed. Mikael Carleheden and Michael Hviid Jakobsen. Burlington, VT: Ashgate, 225–40.

Dalal-Clayton, Barry and Barry Sadler 2014. *Sustainability Appraisal: A Sourcebook and Reference Guide to International Experience.* Abingdon; New York: Routledge.

Dauvergne, Peter and Jane Lister 2013. *Eco-Business: A Big-Brand Take-over of Sustainability.* Cambridge, MA: MIT Press.

Dulcinéia Catadora n.d.. 'About Us'. www.dulcineiacatadora.com.br/sobre__about_3.html. Accessed 16 April 2015.

Dulcinéia Catadora 2012. *Catador (Waste picker)*. São Paulo: Dulcinéia Catadora.

Fundación Mexicana para el Fomento de la Lectura (Mexican Foundation for the Promotion of Reading) 2012. *Encuesta Nacional de Lectura (ENL –National Reading Survey)*. Mexico: Fundación Mexicana para el Fomento de la Lectura (Mexican Foundation for the Promotion of Reading). Mexico. www.educacionyculturaaz.com/wp-content/uploads/2013/04/ENL_2012.pdf. Accessed 15 February 2017.

Guha, Ramachandra and Joan Martinez-Alier 1997. *Varieties of Environmentalism: Essays North and South*. Abingdon: Earthscan.

Hansmann, Ralph, Harald A. Mieg and Peter Frischknecht 2012. 'Principal Sustainability Components: Empirical Analysis of Synergies between the Three Pillars of Sustainability', *International Journal of Sustainable Development and World Ecology*, 19 (5): 451–9.

IADB (Inter-American Development Bank) 2014. *Regional Initiative for the Inclusion of Recyclers*. www.iadb.org/en/topics/solid-waste/regional-initiative-for-the-inclusion-of-recyclers,4918.html. Accessed 16 April 2015.

Medina, Martin 2005. 'Scavenger Cooperatives in Developing Countries'. Paper presented at WIEGO/Cornell/SEWA Conference on Membership-based Organizations of the Poor, Ahmedabad, India, January.

Meza, Aurelio 2014. 'Editoriales cartoneras en México' ('Cardboard Publishers in Mexico'), *Radiador Magazine* 34. www.radiadormagazine.com/2014/04/editoriales-cartoneras-en-mexico.html. Accessed 15 February 2017.

Mocarzel, Evaldo 2011, *In the Margins of Waste*. https://www.youtube.com/watch?v=skhRBIMDEbQ. Accessed 15 February 2017.

IBGE (Instituto Brasileiro de Geografia e Estatística) 2008. *Pesquisa Nacional de Saneamento Básico (PSNB)*. Rio de Janeiro: IBGE. www.ibge.gov.br/home/estatistica/populacao/condicaodevida/pnsb2008/. Accessed 15 February 2017.

Swarr, Thomas E. et al. 2011. 'Environmental Life-cycle Costing: A Code of Practice', *International Journal of Life Cycle Assessment* 16 (5): 389–91.

Tuana, Nancy 2008. 'Viscous Porosity: Witnessing Katrina'. In *Material Feminisms*. Ed. Stacey Alaimo and Susan Hekman. Bloomington and Indianapolis: Indiana University Press, 188–213.

United Cities and Local Governments 2008. 'Agenda 21 for Culture'. Barcelona: UCLG www.agenda21culture.net/index.php/documents/agenda-21-for-culture. Accessed 15 February 2017.

United Nations 1992. 'Agenda 21'. United Nations Conference on Environment and Development, Rio de Janerio, Brazil. http://sustainabledevelopment.un.org/content/documents/Agenda21.pdf. Accessed 15 February 2017.

United Nations 2002. *Plan of Implementation of the World Summit on Sustainable Development*. (Johannesburg Plan of Implementation – JPOI) www.un.org/esa/sustdev/documents/WSSD_POI_PD/English/WSSD_PlanImpl.pdf. Accessed 15 February 2017.

United Nations 2010. 'Sustainable Development'. www.un.org/en/ga/president/65/issues/sustdev.shtml. Accessed 15 February 2017.

World Commission on Environment and Development 1987. *Our Common Future* (Brundtland Report) Oxford: Oxford University Press.

5

Sustainability after extinction: on last animals and future bison

Joshua Schuster

It has been well documented recently that there is a noticeable rise in the rate of extinction across all plant and animal kingdoms. Several conservation biologists have indicated that current extinction rates are now between 100 and 1,000 times expected or background rates of extinction.[1] The rise of extinction rates in the past few hundred years can be situated in parallel to the rise of scientific knowledge concerning the history of extinctions that stretch from recent times to the earliest flourishing of life 3.5 billion years ago. We know more than ever now about the history of life and death on the planet, but we are still witnessing a dramatic loss of species, largely due to human causes that include expansive pesticide use, habitat destruction and over-hunting and fishing. We are becoming terribly knowledgeable about the disappearance of life as it disappears. We are making a science and a culture out of coming to terms with extinction.

But this essay is not about why animals become extinct, nor is it about research done to document and catalogue the number of disappeared species. Rather, my focus here will be on understanding the lives of animals as they approach thresholds of unsustainable population sizes, which calls forth intense anxieties about their imminent loss, but also proves to be controversial grounds for maintaining animals in a diminished state after population collapse. Before animals reach the point of extinction, they often dwell in a prolonged, unsettled status of very small populations. An increasing extent of the world's biodiversity is heading in this direction of diminishing numbers. Alongside the increasing knowledge and care concerning lost animals, we are expanding rapidly the category of last animals – animals that are endangered or nearing extinction. This essay then is about how this nature-culture of last animals has developed, and what this condition means broadly for thinking about current tensions between concepts in animal studies and sustainability studies. Each plant and animal has a minimum number of members it must maintain, a minimum viable population threshold (for most animals, this means numbers between 250 and 500), which demarcates the final line between sustainability and collapse of the species (Soulé 1987). While we know that the overall extinction rate is rising, there is a related phenomenon of many species approaching unviable populations. But because we live in a culture increasingly knowledgeable and self-aware about both extinction and biodiversity, there is now an intensifying scrutiny of how to stall or defer these last stages of animal life.

Indeed, there is so much attention to this critical stage of low animal numbers, fostered by conservation scientists, NGOs and local and international advocates, that it now seems that an endangered animal is just as likely to become extinct as it is to end up being saved from the precipice, only to remain stuck indefinitely in small population pools. With biodiversity increasingly seen as an economic, aesthetic, nationalistic and ecological good, no one wants to see an animal become extinct on one's watch. Yet, conservation budgets and public interest in protecting animal others can only be stretched so far. So, in order to forestall falling below minimal viable populations, one sees an expanding condition of minimal populations as such. The lives of last animals are now largely stuck in a holding pattern where the animal is maintained in small numbers that are deemed minimally sustainable but not much more expandable.

The stages of animal population decline as listed by the International Union for Conservation of Nature (IUCN) is just one way of accounting for the lives of last animals. The categories – least concern, near threatened, vulnerable, endangered, critically endangered, extinct in the wild, and extinct – index numerical crises but also each imply a culturally specific and even ontologically changed conditions.[2] Another example can be found in the proliferation of 'last' books on animals: *The Last Tiger, The Last Lions, The Last Panda, The Last Polar Bear, The Last of the Curlews, Last Chance to See, Last Animals at the Zoo.*

The depletion of animal populations has become such a global phenomenon that, as the journalist J. B. MacKinnon puts it, 'We live in a 10 Percent World' (2013: 38). To a striking degree, animal populations worldwide have been diminished to numbers that total 10 per cent or less of historical populations. MacKinnon cites how the world's biggest fish, including tuna, cod and sharks, have been reduced to 10 per cent of abundance compared to levels in the recent past. A similar percentage of depletions applies to nearly all mammals across the globe, and most bird populations have also seen precipitous declines. Indeed, in many cases, 10 per cent would be an optimistic number. Statistics provided in *The Atlas of Endangered Species* note that giant tortoises in the Galapagos went from 250,000 in the 1500s to 15,000 today; chimpanzees numbered 2 million as recently as 1900 but now number 150,000; the population of all tiger species was estimated around 100,000 in 1900 and now only 2–3,000 remain in the wild (quoted in Mackay 2009). A stunning report issued by World Wide Fund for Nature (2014) calculated that up to 52 per cent of wild vertebrate animals have been lost since the 1970s. Not only population numbers but also the historical range of habitat for the vast majority of animals has been dramatically reduced. Tigers used to roam throughout nearly all of Asia and the Middle East, but are now found only in small pockets of territory, most notably in India, China and Russia, less than 10 per cent of their historical range. The wolf and grizzly bear used to range across almost all regions in Europe, Northern Asia and North America; almost no bears remain in Europe and the wolf has been eradicated in nearly all of its former territory in Europe and much of North America.

How do we understand the condition of animals now that low populations and drastically diminished habitat ranges are the new norms for much of the nonhuman life on the planet? What does the attempt to sustain low numbers of animals mean for both the lives of animals and the concept of sustainability? How do numbers of animals play a role in animal biopolitics, which in the vein of Foucault includes management aimed at the level of population and terms of life for a species? Ultimately, as we will see, the number of an animal plays a key role in the very ontology of the living.

For now, a series of paradoxes and tensions have been set into the landscape. In many cases, animals are allowed to thrive in protected zones but are practically powerless outside these parks and preserves. Indeed, for many animals it is deemed a success that they are able to be maintained in small but stable populations inside conservation enclaves. Sustainability for animals now mostly means existing in small, delimited and scrutinised habitats indefinitely. In this case, sustainability for humans and animals is practically inverted: humanity defines sustainability as a way to support consistently a peak population of around 7–9 billion, while for many animals, 10 per cent of peak population numbers would be seen as a triumph. Human sustainability is pitched at the level of the whole planet, while for most animals sustainability has shrunk to a very small area of historical habitat range. Sustainability for humans may mean limiting economic growth, while for most animals it means finding some way to stave off complete elimination.

The language of sustainability, beginning in its early adoption in the United Nations Brundtland Report, *Our Common Future* (1987), loudly tied anthropocentricism to developmentalism: 'Humanity has the ability to make development sustainable to ensure that it meets the needs of the present without compromising the ability of future generations to meet their own needs' (World Commission 1987: 41). The sustainability of animal lives and species populations barely warranted mention in the report. The report raised alarms about increasing extinctions, but couched these in the context of narrow declarations of self-interest, stating that we should be concerned about biodiversity loss because 'Species and natural ecosystems make many important contributions to human welfare' (World Commission 1987: 125). Subsequent to Brundtland, biological enclaves and hotspots have been imagined to be happy mediums for both animals and humans: here biodiversity can be concentrated for animal sustainability, while the rest of the planet is colonised for human sustainability.[3] These enclaves are easier to visit because they are compact destinations and therefore support ecotourism and are more amenable to patrol against poachers. Condensed biodiversity hotspots, then, are able to make money to maintain themselves and provide for human economies near these areas. Animals agree somehow to remain in small numbers in their downsized ranges, while humans agree to leave them more or less unimpeded while paying for the privilege to observe contemporary ecology in action.

These enclaves (Myers et al. 2000: 853), which now largely describe the *de facto* habitats of the 10 per cent, offer a special kind of animal capital. Inside these zones, economics and animal population size reach a kind of collusion. For example, lions used to roam almost all of Africa, the Middle East, Asia and even much of Europe. Now that 40,000 or so lions remain, they can be kept in smaller, concentrated areas, which are easier to visit, manage and keep separate from humans. A limited number of permits can be allowed for hunters willing to pay up to $30,000 (the current price to hunt a lion in Botswana), which provides the funds for further lion conservation and sustains local economies, which include the humans who have to face lion attacks from time to time and need a reason not simply to shoot the animals. If the lion population were larger, the permits would not be worth as much, there would be more attacks on humans, and, without the shadow of extinction looming, conservation groups would be less inclined to spend time and money on the animal's behalf. Last animals receive huge investments in research, conservation and public attention, but when numbers rebound, these resources often shift elsewhere according to a 'triage' logic, which in turn tends to result in the animals returning to lower population numbers as conflicts with humans recommence. Paradoxically, fewer animal numbers may be the pathway to more biodiversity preservation in general.

Outside of these enclaves, there seems to be no reliable formula for human–animal co-existence for most of these last animals today, even though as recently as just a few generations ago, humans regularly encountered these animals in their range and abundance across the globe. MacKinnon adapts a term from psychology to describe this phenomenon as 'shifting baseline syndrome' (2013: 16), in that the depletion of animals in number and range becomes the new norm or baseline for the next generation. Each generation thereby encounters fewer animals that occupy smaller ranges, and assumes that the current lay of the land represents the standard population size and habitat for the animals. This is one way MacKinnon attempts to explain how most humans have become used to seeing a 10 world world as nothing all that alarming – they have lost the memory of past abundances, having only grown up with vastly diminished animal populations around themselves.[4]

A 'shifting baseline' makes the agency behind these shifts seem vaguely dispersed, if not intentional. It may be rather that many, perhaps most, humans have wanted a 10% world, especially in regards to carnivorous animals that would eat us or our animal property, or herbivores that eat our plant property. A 10 per cent world right now is very comfortable and convenient – more or less what we like to define as sustainable. In most parts of the world, we see animals when we want to and keep them apart at all other times, for their safety and ours. To reveal the brutality at the heart of this landscape of comfort, MacKinnon cites a phrase from the biologist Norman Myers, who called the rapid disappearance of animals in the past few centuries 'the great dying' (Myers et al. 2000: 35). But this is also the same period as the 'great acceleration' for humans and the development of technologies. Furthermore, there now are more protected parks and conservation areas than ever before. Hence one might suggest that animals have traded population numbers for taxonomic protection. Even though extinction rates are rising, a large portion of Earth's biodiversity hotspots remain intact for the most part due to carefully planned conservation efforts, suggesting that it is possible to have a world full of species without being full of animals. In many cases, the planet remains biodiverse without much biomass.

Of course, the notion that animal populations will stabilise somehow at 10 per cent in plush, well-demarcated enclaves is another convenient fiction. Animal populations may continue to be depleted even if humans decide to stop bothering an animal once it passed into 10 per cent of population and range, as disruptions could be caused by other factors, including breakdowns in the food chain, disrupted migration corridors and newly introduced predators and competitors. It might also be the case that the genomes as well as the behaviour of animals change at these low levels – some crucial gene variations may have been lost, and learned hunting, navigating or reproduction skills may not be passed on consistently in smaller groups (Soulé and Scott Mills 2014: 189–211). It could even be the case that certain animals could not be scaled back up and returned to previous places given the changed genome or behaviour. The historical range of the animal might already been changed irreversibly into a new landscape; the animal might be a different animal now too. Also, enclaves are not as protected as one would hope, as habitat fragmentation has its own problems, while an enclave is also more vulnerable to suffer an extreme environmental event (burst of cold/hot weather, a massive toxin spill or a harmful insect or bacterial outbreak) that causes localised ecological collapse in a preserve that was thought stabilised and sustainable. Down in the 10 per cent, squalls make for shaky conditions, ecosystems teeter this way and that, leaving windswept animals unsure of what is ahead and to what extent the past will be of any guide to the future.

How the buffalo roam

At this point, I would like to examine the above issues with an eye to a specific historical example of how life is lived below the 10 per cent: the near-extinction of the North American plains bison (*bison bison*) in the late nineteenth century, and the subsequent attempt to save the last remaining animals, and then to repopulate them in their former range of habitat in the twentieth century. I will only present the historical details of this event in brief outline, because I am interested primarily here in the broader cultural and scientific impact this animal has had on thinking the possible afterlives of animals in the wake of an extinction event, where the choices for the bison were eradication, remaining indefinitely in small numbers, or large-scale repopulation. Arguments for this last option have recently been gaining momentum, using the rhetoric of sustainability in support. A series of manifesto-like claims for repopulation of the bison appear in Ernest Callenbach's *Bring Back the Buffalo! A Sustainable Future for America's Great Plains* (1996). Callenbach, previously famed for *Ecotopia* (1975), declares in the title of his book his aim to hitch the return of the bison to its previous population size to a full-blown project for long-term human–animal sustainability in the North American plains.

The bison, also commonly called buffalo (although this is now considered a misnomer as the bison is not in the same genus as the African or Asian buffalo), is estimated to have numbered at least 30 million (and possibly as many as 60 million) at its peak population, which likely persisted until the early 1700s. Bison habitat once stretched across plains from Texas to northern Alberta, and from the Rockies to the Atlantic, but by the late eighteenth century, the animal was found only to the west of the Mississippi. Hunting by Native Americans over thousands of years had not much changed the overall animal numbers, while the bison provided Plains Indians with a bounty beyond meat, including bones used for tools, sinews for sewing, pelts used for clothing and housing, and bladders used to carry water. Interest in bison hunting by white settlers had been relatively limited until after the American Civil War concluded in 1865. The end of the war left a wave of men unemployed and skilled in shooting newly improved rifles. With freshly laid train tracks depositing hunters in the plains, teams of hunters and skinners had formed to pursue the bison in systematic fashion during their herding season in the autumn. Each group of hunters was capable of killing a few thousand animals a year, stripping the animals of their hides and tongues, leaving the rest to rot. Several thousand of these teams covered the plains, shipping to the east by train hundreds of thousands robes per year. By the mid-1870s, the bison had been hunted to extirpation in the southern area of its range; in the northern area, the hunt continued up to 1883, collapsing the bison population entirely (Dary 1974; Isenberg 2000).

This extinction event gathered together many key inventions of modernity: fast-loading rifles with longer reach, convenient train depots, manufacturers in the east who needed tougher leather for complex belting systems, a popular press that documented the killings with articles, letters and cartoons, a rising commodities financial and corporate network, and ranchers and homesteaders ready to move into cleared areas and use new farming technology to get the plains to grow food. With this technological and social convergence aimed at harvesting the bison, it took about two decades to go from 30 million to approximately 1,000 bison left on the planet. These few animals remained in scattered pockets that included the protected area of Yellowstone National Park and the New York Zoological Society's Bronx Zoo, as well as being kept in small numbers on a few private ranches that held onto the animals.

Even as the animal was massacred *en masse* in the 1870s, a number of attempts to safeguard the bison were launched by members of animal protection societies, concerned ranchers, those sympathetic with the plight of Native Americans and those who protested against the notion of rendering an animal extinct on principle. On the other side, there were many who applauded the slaughter as a way to debilitate the remaining Plains Indians who skirmished periodically with frontier settlers. Quickly the animal took on different nationalistic characteristics: it was identified as a casualty of manifest destiny, as well as associated with the figure of the primitivised noble savage, a hardened, powerful survivor of a land with few trees, little rain and poor soil. Hunters and ranchers saw the animal as an object of American opportunity, first by rendering its body into capital, then later as useful in its absence for giving way to massive cattle ranching and monocultural farming. Its death, like the sweeping away of so much other life, was seen as necessary and inevitable for American expansionism. But also, arguably for the first time ever in the context of natural history, a burgeoning class of natural scientists and conservationists took a stand against the defeat of the animal, with the recent knowledge of the biology of extinction as added motivation. Led by William Hornaday, who established a small preserve for the animal in the National Zoo in Washington DC in the late 1880s, the American Bison Society formed to launch an organised preservation of the animal. Hornaday's group worked to track the remaining animals, manage a small herd, pass legislation that established no-hunting reserves, and develop research methods that would aid in growing the population back from such radically small numbers. The group achieved dramatic success in the first intentional attempt to use conservation science to prevent extinction, with really no margin for error.[5] By the 1930s, the bison population had grown to over 25,000 in number, spread in wild herds and some on ranches and in zoos, and the group declared their work a success and effectively disbanded. Within a few decades, the population would continue to grow, plateauing to over 300,000 by 2000, but most of the population increase would be on private ranches, with the animal destined for consumption.

Today, bison numbers are still well below 10 per cent of historic populations (actually they are much closer to 1 per cent – and of those remaining numbers, only about 10 per cent live on land protected from hunting and ranching), while most conservationists have moved on to concern for other endangered animals. But in the 1990s, the bison again caught the imagination of conservationists who wanted to propose something big, bold and captivating for the newly hatched idea of 'rewilding', which involved reintroducing animals to past habitats in order to reconstruct fragmented and depleted ecosystems. Strong calls for various rewilding projects have been made by Gary Snyder (1990), Paul Martin (2005), Dave Foreman (2004), Emma Marris (2011), George Monbiot (2014) and Marc Bekoff (2014), among others. Rewilding is not just about bringing the animals physically back, but also includes advocating for a spiritual and cultural reclamation associated with the return of specific regional species and ecosystems. Further implied in rewilding is the notion that there would be a kind of literary and aesthetic reawakening with the return of such animals, involving a much-sought overlap of textuality and wildness.

Consider, for example, Snyder's poem 'Home on the Range', which appropriates the nationalist nostalgia of the American folk song to envision a wholly bison-enabled way of life. The poem begins: 'Bison rumble-belly / Bison shag coat / Bison sniffing bison body' (Snyder 1992: 356). The actual presence of bison on the range may be severely diminished, but this poem circumvents the missing animal by bringing the name of the bison up close and activating all the senses of both animal and reader. There is no subject, no 'I' to mediate experience, just the directness of the bison that can be heard, touched, smelled, tasted and everywhere seen. The most comprehensive rewilding would even change the power and purpose of our senses. Reader and bison seem to be in the same intimate space, as the poem does not distinguish between the 'liver warm' inside the bison and the same object being consumed by a human. The only thing that indicates mediation in the poem is the quotes around the folk song 'Home on the Range', a cue to the reader that this song and title is being rewilded too. Certainly wilding the imagination is an important step in dislocating the imaginary from its settled ways, yet the transitions from textual rewilding to large-scale rewilding projects in widely developed landscapes requires close attention for the kinds of claims made upon animality and sustainability together.

Callenbach's *Bring Back the Buffalo!* – with its exclamation point usage akin to the eco-defence group Earth First! – positions itself between speculative imagination and practical planning as it takes up the cause to repopulate the animal across the plains. Callenbach saw this project as a convergence of idealist activism and pragmatic conservation as a model for future ecological movements. The rewilding of the bison offered Callenbach a post-utopian agenda which would promise practical solutions that still held big ecological payoff.

The shift from a literary utopia to an actual landscape for Callenbach also meant rethinking the lives of animals from his previous work. Looking backwards at Callenbach's *Ecotopia* (1975) from the position of animal issues today, one notices immediately the problematic, simplistic focus on human-centred, ego-psychology portrayals of the urban–rural relationship. Across this utopian landscape one finds a curious lack of animals and animal activism. Early in the novel the journalist narrator William Weston encounters a group in downtown San Francisco coming back from a deer hunt with the animal carcass, and interprets this encounter as a 'throwback policy' (Callenbach 1975: 16) a prototype for a version of today's caveman diet movement. 'Ecotopians, both male and female, have a secure sense of themselves as animals', so the reader is told (Callenbach 1975: 32). The novel is chock-full of examples of humans wanting to feel wild and loose, where the animal is read as sexual, spontaneous and meaty. At the same time, ecotopians have economic, social and ecological systems running at a purr. Human life is secured as animal, but ironically, animal life is made even more insecure by its role as 'absent referent', in Carol Adams's phrase (2010: 66). Citizens of ecotopia try to live a sentimentalised frontier lifestyle, happily playing cowboys and Indians at times, in combination with planned community principles and 'biological abundance' (Adams 2010: 65). The heterosexual politics of *Ecotopia* are deemed to be as important as the ecological politics, implying that a heteronormative libido would assume also a libido for greenness. The rhetoric of the wild is more frequently attached to women in the novel (especially Marissa, the narrator's love interest) than any other entity.

One of the rare moments apart from this anthropocentric utopia is the description of a 'steady-state' systems ecology seen from the point of view of a mouse. The journalist visiting Ecotopia is told that, from a wide lens view, ecosystems look stable and static, but Callenbach has one Ecotopian explain that there is plenty of niche dynamism, such as mice eating seeds and hawks eating mice. 'I begin to see what you mean', says Weston. 'It may not look so static to the mouse' (Callenbach 1975: 34). But it is not that the life and cares of this particular mouse are of much importance, and the conflicts between human interests and animal interests apparently melt away. The lush steadiness of the state of things, rather than biodiversity preservation or animal welfare, becomes the standard by which to measure sustainable modes of human dwelling. Ecotopia evidently superseded animal utopia. Anything redolent of animal liberation, promoted especially by Peter Singer in his book published in the same year as *Ecotopia*, is nowhere to be found. Moreover, the notion that a properly ecological community had to be stable and predictable made for a comforting fantasy in the face of rising extinction rates, biodiversity collapse, petroleum shocks and resource wars. This homeostatic ecological paradigm was further undermined with the introduction of chaos theory into ecology science in the late 1970s. Suddenly *Ecotopia* looked to be way too much about blissful eco-minded consumerism, social consensus and homogeneous self-organisation (remember that in the novel all the black people in California decided, of their own spontaneous accord it would seem, to relocate to 'Soul City'), as if these were the real ecotopian values. There are apparently no problems with any biological hazards such as invasive species, extinctions, pathogens or large-scale, confined animal industries, since ecology has been preset for steady-as-she-goes. Despite the celebration of the unwashed, waste-free hippy lifestyle as the greatest good, *Ecotopia* turns out to be way too clean, managed, heteronormative, pain-free and quiet (no electric guitars, please, and God help us from our unwashed rivals, the punks) to have to bother with including forms of otherness that would not abide by the steadiness consensus.

In *Bring Back the Buffalo!*, Callenbach ostensibly puts animals front and centre as a corrective in this vision for the future of ecology but, as we will see, many of the same problems of *Ecotopia* return, albeit in changed form. This book calls for a future ecology to be focused on the sustainable rather than the utopian, and Callenbach defines sustainable as 'an ecosystem that can endure stably over a long period' (1996: 4), at least several thousand years. The steps to bison repopulation outlined by Callenbach are not in themselves very radical or demanding of a whole new political outlook. The plan would be to coordinate government and private land purchasers into obtaining a wide swathe of the plains, land that is not particularly arable without a tremendous input of fertilisers, petroleum-based machines, excessive aquifer usage, and reliance on other subsidies. The huge tax revenues allotted by the US government to farmers, whether or not they grow, instead could be used to buy land back from farmers, presumably saving money in the long run and stemming the tendency of the government to prop up idle and unprofitable farms. The purchases would be strategic, with the idea of creating corridors between national parks and following the historical paths of bison migrations. Here 'we must learn to "think like bison"' in land usage (Callenbach 1996: 29) so that bison and grassland develop with each other. Bison chew only the tops of native grasses, letting them regrow, while cattle, currently numbering over 110 million in the US, rip grass from the roots (due to the cost of feeding them native grasses, cattle today are largely fed corn and other grains as substitutes).

Callenbach projects that several million bison could be stocked on these lands, needing very little hands-on attention or fencing, as they live peacefully enough among human developments if left alone (people would be need to be taught to give the animals a wide berth because bison may attack if approached). The bison variously could be public or privately owned, or under the auspices of Native Americans, or in nature conservancy preserves or large ranches. Here is where Callenbach tries to piece together a version of sustainability in the contemporary economic parlance, as exemplified by Paul Hawken, Amory Lovins, and L. Hunter Lovins's *Natural Capitalism* (1999), where private enterprise and environmental benefit are seen as synergistic. Most controversially, Callenbach is ready and willing to see these animals as fully absorbed into capitalist systems of surplus value and the industrial animal-rendering empire. Effectively all the millions of bison would be available for rendering and consumption, and Callenbach envisions Native Americans and white settler ranchers having free run of the killing of these animals as long as they do it to maximise food and keep the population of the herd restocked.

This manifesto for sustainability on the plains is not a manifesto for animal rights or independent animal flourishing, and sustainability here really means, once again for Callenbach, a romanticised ecological stability that supports a slightly curtailed standard of human consumption. The bison would return in huge numbers, but their increased population would be a tacit consent to be animal capital in a world of highly organised ranchers, slaughterhouses and meat retailers. Callenbach supplies many pages of commentary extolling how tasty the bison is, how healthy and low fat the animal is for the hearty meat-eater, and how promising the animal would be in the hands of fast food operators. With no hint of irony, Callenbach suggests a 'McBuff' burger (1996: 197) as the next great food item. He happily details how to buy bison meat and have it shipped to your door, and he genially supplies recipes at the end of the book. His book is full of 'bison entrepreneurs' (Callenbach 1996: 190), from media mogul Ted Turner to small, one-person ranches, from big beef industries to mom-and-pop bison burger joints. Callenbach quotes Harold Danz, executive director of the National Bison Association, who states unabashedly that, 'To preserve buffalo, the best thing we can do is eat them. Animals that people eat do not become extinct' (1996: 186). Lest this shock the sheepish animal welfare reader, Callenbach offers a quote from the trusted environmentalist poet Gary Snyder on the next page, who says with equanimity, 'There is no death that is not somebody's food, no life that is not somebody's death ... Eating is a sacrament' (Callenbach 1996: 187). Along the lines of what Thom van Dooren calls 'killing for conservation' (2011: 286), here we have eating for conservation. Consuming and rendering is declared necessary in order to rewild. The logic of sacrifice and sacrament is repurposed to make animal numbers grow. In the great American fantasy, eating more is the pathway to more abundance, while death is ever fecund.

It is fairly clear already that Callenbach hitches the notion of sustainable bison ranching to a discourse of settler nationalism that, he assumes, will cross-fertilise each other. Indeed, we are told from the outset, 'the fate of the bison may well prove emblematic of the future of our nation' (Callenbach 1996: 1). It helps that the bison is easily romanticised according to well-trodden tropes of America's nostalgia for itself. 'Strength, endurance, adaptability, and cooperation in the face of danger make the bison a striking emblem of America' (Callenbach 1996: 2). The masculine, no-nonsense, American football-sounding, cowboy rhetoric used here is something even a Reagan-ite could love (and appears as a patriarchal reversion in comparison to the examples of female empowerment in *Ecotopia*). 'Bison are quintessentially American animals: stalwart, noble symbols of wildness, freedom, and self-sufficiency' (Callenbach 1996: 9). And if they weren't? What does this kind of nationalistic conservation rhetoric say about the sustainability of other animals? Yet Callenbach only declares loudly what other rewilding proponents tacitly condone: harnessing nationalism and animal nativism appears to be an effective way of repopulating animals, since these two agendas are likely to prove supportive of each other. But one has to wonder if living in a world of repopulated animals would mean a return to biopolitical nationalism that eagerly employs animals as models for military strength (as one sees already with the long-standing use of animal names for military hardware). Also, in Callenbach's version of rewilding, the renationalised bison once again becomes a political animal, but not an autonomous citizen. Debates over animal rights cannot be reopened; this would spook away conservatives who might otherwise be sympathetic to the cause of repopulation. This 'buffalo commons' (Callenbach 1996: 199) is still a variation of enclosure.

At the same time, there is nothing intrinsically wrong with envisioning animals as noble and desirable to repopulate based on an affective and cultural longings. Callenbach is right to say that the bison's 'absence is our loss, psychologically, spiritually, and morally' (1996: 16). He also adds intriguingly that the bison, which can easily weigh 2,000 pounds, 'is the only large wild animal with whom there is any prospect of sustained coexistence on mass term' (Callenbach 1996: 16). (Some have even argued that North American ecosystems miss the mammoth and camel, and these too should be rewilded, with elephants as substitutes for mammoths). As in *Ecotopia*, if you look at the big picture, it is very satisfying. 'On the Plains we can transform current petroleum-based farming and ranching into an enduring, self-reliant system resisting on the perennial resources of the region: sun, grass, and wind. On the Plains, a deep planetary challenge of long-term human survival waits to be met'(Callenbach 1996: 3). Herds can graze under giant windmills, people and bison can live together, and new technology can converge with the ancient, prosperous ways of animal–human co-existence. Perhaps even what it means to be American will change.

> If we decide that it is fitting for these noble beasts to share our future, and make room for them on the continent again, we will be a different people. It is worth entertaining the possibility that we will be a more humble, less driven, less exploitative people, with a livelier sense of connection to the wild in ourselves as well as in bison. (Callenbach 1996: 258)

Rewilding is always this moral as well as biological project, intertwining species and spirit, yet both these aspects require imaginative and critical analysis. Rewilding aspires to bring back animal numbers and animal freedoms in order to let ecosystems be, even while welcoming large-scale conservation management and monitoring practices that constantly change what wildness means. To rewild means to manage the landscape without appearing to do so, buffing up the reappearance of the frontier with scientific approaches, nationalist projects and media-savvy displays in mind.

The post-sustainable bison

'Americans love happy endings, and the story of bison puts one within our grasp', Callenbach proclaims (1996: 149). In recent decades, advocates of the managerial kind of sustainability have sought to incorporate such positive and heart-warming messages, to varying effect and criticism. It is not that negative or apocalyptic warnings would improve things for the bison, but instead of promising 'happy endings', more honesty and awareness about the history, politics and ecological reality of the animal would be a gain for all involved. Dave Foreman repeatedly points to acknowledging 'wounds' in the landscape as a necessary step to rewilding (2004: 3). Callenbach is certainly aware of the longer traumatic history of white settler treatment of bison and the Native Americans dependent on the animal, but he acutely avoids any emphasis on this brutal legacy and offers little evidence that the traumas of the animal informs his current thinking of how to rewild it. The palpability of the return of bison has become a recurrent theme in essays and cultural works by many Native American and Canadian First Nations peoples in recent years. In comparison to Callenbach's boosterism, what stands out noticeably in indigenous artworks supporting bison repopulation is the importance of thinking through trauma, a critique of American nationalism and desire for the animal at the same time. Winona LaDuke has written of the need to connect the history of radically curtailing where the bison roam and the many cases of US government reduction of the size of reservation land previously allotted to Native American tribes (1999: 139–66). The figure of the bison depicted in literature and visual works by recent First Nations artists – for example Thomas King's novel *Truth and Bright Water* (1999) and Tasha Hubbard's short animated film *Buffalo Calling* (2013) – draw attention to how the return of the animal links past and present in a landscape that is tense, violent and politicised, but also musky, drenched in symbolism and ripe for new human–animal relations. Further examples of a vision of the fraught renewal of the bison can be found in drawings by Adrian Stimson, Blackfoot, and paintings by Kent Monkmon, Cree, as both see the bison in its post-extinction phase caught between a haunting and passionate existence.

Stimson's series of acrylic paintings, *Bison Fission* (2010) and *Fuse* (2010) (Figure 5.1) show bison grazing in front of a nuclear explosion, each image depicting the mushroom cloud in a different phase. The bison appear to stroll and munch calmly as the ultraviolent nuclear fission bomb detonates in the near-distant horizon. The quiet gathering of the animals is engulfed by the explosion that exceeds the scales of sound. In this atomic flash, we see a bison-military-energy complex. Whereas Callenbach also saw an energy–bison connection in which one would sustain the other, in Stimson's works the viewer is about to see both the landscape and the bison wiped out. The two series stage scenes of annihilation but also a recognition of how bison lives and Native American lands form

Figure 5.1 Adrian Stimson, *Fuse 2* (2010)

the larger backdrop to the black-and-white colours of early nuclear test photos (although almost all continental nuclear tests were in Nevada, where bison did not roam). These paintings do bring the bison back, but show how the landscape is no longer the same in a post-nuclear age. By capturing the split-second time of the mushroom cloud emerging, we see juxtaposed the placid repose of the bison and the radical violence and militarisation visited on the land. Instead of a rewilding that longs to turn back the clock of history, we get a sense how both animals and nuclear science have contributed to the current ecology of the southwest and plains.

In another visualisation of bringing back the buffalo, Monkman's oil painting *The Chase* (2014) (Figure 5.2) shows a herd of bison and a few bulls stampeding through a non-descript, run-down city street. The animals are pursued by an American Indian hunter, riding a rocket motorcycle, clad in sexy boots and carrying a Louis Vuitton-inspired arrow holder. Most of the running bison and bulls are depicted naturalistically, but some are cubist and some are in the style of paleolithic cave art. In some unknown city, an unusual combination of animal, technology, sex, art, primitivism, queerness and Native cultural revival come together in this image. The bison is brought back not in support of a nationalism–sustainability consensus, but as a queer object crashing through the streets, pursued by a transgender hunter-artist. Monkman – who has painted many scenes of bison in the setting of nineteenth-century-styled North American landscapes, often incorporating subtle expressions of non-heteronormative sexual activity

Figure 5.2 Kent Monkman, *The Chase* (2014)

– here depicts the bison as spilling into the city and bringing together 30,000 years of bison art. The return of the bison today shows the animal embedded in a world caught up in fetishising the animal body, comfortable with seeing the bison amidst other luxury goods and sensual consumer items. With bison running in paved streets and an Indian shooting a bow while riding a motorcycle, everything seems a bit out of place and time, yet the painting embraces living in a world full of anachronisms, inviting rewilding to add more. But if the bison were to be brought back today, they would be joining a world where queerness and wildness mix, rather than entering the streamlined masculine bison–rancher bonds that Callenbach fantasised.[6]

These contemporary works do not foretell a world where rewilding would be outright dismissed, but neither do they suggest that animals will come back or promise a 'happy ending' where energy, meat and nationalistic unity will be forever bountiful. Instead, these artists connect an intense longing for the wildness of the bison as inseparable from critiques of settler colonialism and the attachment of the animal to some of the utmost forms of violence imaginable. Bringing back the bison also means bringing back ghosts, whatever their nature. Furthermore, the point of bringing back the bison is not to institute a managerial form of bison capitalism, but to queer sustainability itself in order to welcome new and various ways to desire and live with the animal. An indigenous-led return of the bison could certainly be sustainable as Callenbach defines it, but the return of the animal comes in the context of a changing, ongoing relation of Native Americans to indigeneity. After a deeply traumatic extinction event, the bison does not return simply as before, rather the history of that extinction event remains in the very being of the animal. To borrow a term from Allan Stoekl, the bison is 'post-sustainable' (2007), coming after the illusory dream of steady-states and closed environmental circuits that disregard ecological realities of loss, disruption, extinction, excess, joy, violence, queerness and waste. The masculine, nationalist project of rewilding instead would become more diversely wilded in this queering of bison, sustainability and indigeneity together. It would mean imagining a future for animals without necessarily tying them to some version of animal capital. Ecotopia cannot circumvent ecological trauma, and sustainability does not erase extinction, but creates amidst it. The desire for the return of the bison, and a wilder biodiversity, takes us into uncharted pungent ecologies and musky ways.

Notes

1 In *Extinction Rates*, one of the first collections of statistical analyses of extinction rates, the editors summarised the rise in current extinction rates according

to orders of magnitude. The editors also used the phrase 'sixth mass extinction' (Lawton and May 1995: 20) once, an early example of this now widespread descriptor to account for the collapse in both population size and the rising rate of extinctions of plants and animals. Further discussion of the recent rise of rates of extinction is featured in Richard Leakey and Roger Lewin (1996), E. O. Wilson (2003) and Elizabeth Kolbert (2014).

2 The IUCN primarily uses population numbers to define these categories. For example, 'critically endangered' can mean reduction in population size to 10 per cent of numbers over the last ten years, or a population size of 250 or fewer mature individuals (IUCN 2012).

3 An influential article that appeared in 2000 in *Nature* identified twenty-five primary hotspots that occupy just 1.4 per cent of the land surface of the earth (Myers et al. 2000: 853). The authors of this study argued that focusing on these hotspots, which contain a concentrated diversity of species (but not necessarily supporting large population numbers), would be a more cost-effective way to pursue conservation.

4 The initial emphasis on identifying how 'shifting baselines' become the new normal was developed in 2002 by several marine biologists, divers and film-makers, who set up a media and research campaign at www.shiftingbaselines.org. Further discussion of the issue of massively reduced populations now taken as the norm can be found in Caroline Fraser (2009: 294–9).

5 For a detailed history of the work to repopulate the few remaining bison, see Mark V. Barrow, Jr. (2009).

6 The linking of queer and wild made here owes much to the recent work of Jack Halberstam, who has a forthcoming book on the subject. See also Jack Halberstam (2013).

References

Adams, Carol J. 2010. *The Sexual Politics of Meat: A Feminist-Vegetarian Critical Theory*. New York: Continuum. 20th anniversary edn.

Barrow, Jr., Mark V. 2009. *Nature's Ghosts: Confronting Extinction from the Age of Jefferson to the Age of Ecology*. Chicago: University of Chicago Press.

Bekoff, Marc 2014. *Rewilding our Hearts: Building Pathways of Compassion and Coexistence*. Novato, CA: New World Library.

Callenbach, Ernest 1975. *Ecotopia*. New York: Bantam Books.

Callenbach, Ernest 1996. *Bring Back the Buffalo! A Sustainable Future for America's Great Plains*. Washington, DC: Island Press.

Dary, David A. 1974. *The Buffalo Book: The Full Saga of the American Animal*. Chicago: Sage Books.

Foreman, Dave 2004. *Rewilding North America: A Vision for Conservation in the 21st Century*. Washington, DC: Island Press.

Fraser, Caroline 2009. *Rewilding the World: Dispatches from the Conservation Revolution*. New York: Picador.

Halberstam, Jack 2013. 'Go Gaga: Anarchy, Chaos, and the Wild', *Social Text*. 31 (1): 123–34.

Hawken, Paul, Amory Lovins, L. Hunter Lovins 1999. *Natural Capitalism: Creating the Next Industrial Revolution*. New York: Little, Brown and Company.

Hubbard, Tasha (dir.) 2013. *Buffalo Calling 2013*. Animated short film. [DVD] Canada: Tasha Hubbard Distribution.

Isenberg, Andrew C. 2000. *The Destruction of the Bison: An Environmental History*. Cambridge: Cambridge University Press.

IUCN 2012. *IUCN Red List Categories and Criteria: Version 3.1*. Cambridge: IUCN. 2nd edn.

King, Thomas 1999. *Truth and Bright Water*. Toronto: HarperCollins.

Kolbert, Elizabeth 2014. *The Sixth Extinction: An Unnatural History*. New York: Henry Holt.

LaDuke, Winona 1999. *All our Relations: Native Struggles for Land and Life*. Cambridge, MA: South End Press.

Lawton, John H. and Robert May (eds) 1995. *Extinction Rates*. Oxford: Oxford University Press.

Leakey, Richard and Roger Lewin 1996. *The Sixth Extinction: Patterns of Life and the Future of Humankind*. New York: Anchor Books.

Mackay, Richard 2009. *The Atlas of Endangered Species*. Berkeley: University of California Press.

MacKinnon, J. B. 2013. *The Once and Future World: Nature as it Was, as it Is, as it Could Be*. Toronto: Random House Canada.

Marris, Emma 2011. *Rambunctious Garden: Saving Nature in a Post-Wild World*. New York: Bloomsbury.

Martin, Paul S. 2005. *Twilight of the Mammoths: Ice Age Extinction and the Rewilding of American*. Berkeley: University of California Press.

Monbiot, George 2014. *Feral: Rewilding the Land, Sea, and Human Life*. Chicago: University of Chicago Press.

Myers, Norman et al. 2000. 'Biodiversity Hotspots for Conservation Priorities', *Nature* 403: 853–8.

Snyder, Gary 1990. *The Practice of the Wild*. New York: North Point Press.

Snyder, Gary 1992. *No Nature: New and Selected Poems*. New York: Pantheon Books.

Soulé, Michael E. (ed.) 1987. *Viable Populations for Conservation*. Cambridge: Cambridge University Press.

Soulé, Michael E. and L. Scott Mills 2014. 'Conservation Genetics and Conservation Biology: A Troubled Marriage'. In *Collected Papers of Michael E. Soulé: Early Years in Conservation Biology*. Washington, DC: Island Press.

Stoekl, Allan 2007. *Bataille's Peak: Energy, Religion, and Postsustainability*. Minneapolis: University of Minnesota Press.

van Dooren, Thom 2011. 'Invasive Species in Penguin Worlds: An Ethical Taxonomy of Killing for Conservation', *Conservation & Society* 9 (4): 286–98.

World Commission on Environment and Development 1987. *Our Common Future* (Brundtland Report). Oxford: Oxford University Press.

Wilson, E. O. 2003. *The Future of Life*. New York: Vintage.

World Wildlife Fund 2014. *World Wildlife Fund Living Planet Report 2014: Species and Spaces, People and Places*. Gland: World Wildlife Fund.

6

The twilight of the Anthropocene: sustaining literature

Claire Colebrook

Over the last decade the claims made for the importance of literary understanding, environmental humanities and imaginative reflection have received a (perhaps tragic) reinforcement from the inverse relation between the threats facing humans and other species, and the capacity for action. It is almost as if the prospect of calamity and unprecedented change is so intense that the practical, rational and imaginative resources we have for thinking about the future are simply and woefully inadequate. If reason, calculation, practice and day-to-day thinking have led to a tragic inability to think beyond the expediency of the present, then perhaps it is the task of literature and the humanities to address our sentimental, affective, habitual and non-cognitive comportment towards the world. Scientific knowledge and the dissemination of facts, warnings and already-incurred losses have not only made little impact on effective policy and lifestyle change; the intensity and enormity of the problem may have generated a sense of practical impossibility. Here is where one might turn to the problem of the sublime and deconstruction. Is it possible to accept the inhuman intensity of the problem of the future – its necessary capacity to outstrip calculation and imagination – without abandoning the task or problem of survival altogether? Rather than engineering Nature, the humanities or the imagination in order to ensure 'our' survival, one might ask whether there has been an excess of comprehension in the face of a time and history that has not been paralysing enough. That is, in the face of the failure of scientific know-how to have significant impact on the ways in which 'we' manage our future, perhaps it is *not* literary know-how, expertise and cognitive expansion (and certainly not the environmental humanities) that we ought to embrace in order to sustain ourselves. What I would suggest, perhaps counter-intuitively, is the importance of thinking in terms of a deconstructive or *material* sublime.

As Bruno Latour (2011) has noted, there was a time when Nature performed the role of the inconceivably infinite and therefore sublime force, against which the inconstancy of human life and history seemed insignificant. Now, he argues, it is human history that appears immutable – as though capitalism and the logic of the market were beyond our power – while Nature is changing rapidly. This reversal, he insists, needs to be reversed yet again in order that what appears immutable and beyond our measure returns to the domain of what can be made, and unmade; there is *nothing* natural, immutable or sublime about capital. Perhaps, then, we need to deconstruct sublimity. In its Romantic mode (as Latour reminds us), one could appeal to the infinite, immutable vastness of Nature's 'ample power / To chasten and subdue' (to deploy Wordsworth's language (1985: 37)). But, just as Latour led the way in the notion that Nature does not exist but is always composed, so Wordsworth also posited another presence, 'a sense sublime / Of something far more deeply interfused' (Wordsworth 1936). That is to say, what is sublime is neither nature nor humanity but their impossible and unthinkable imbrication. This is what I would refer to as the material or deconstructive sublime. Before there is a Nature that is infinite, immutable, eternal and that offers 'us' either solace, recompense or moral grandeur, there is inscription: the coming into distinction of living beings, the orientation and inflections that eventually constitute 'a' Nature and a humanity.

One could not, then, simply deploy literary culture to change the way we think, as if there were a 'we' or mindset that could be refashioned or engineered; nor would Nature, with the Anthropocene, finally make itself felt as bound up with our history and therefore worthy of the rights and care we grant to humans. What appears as Nature is, to deploy Latour's terminology, an effect of composition. Who or what composes Nature, Gaia or life is an effect of composition. To confront composition, inscription or materiality is not only to abandon the lure of the 'we' of the new managerialism of the environmental humanities and geoengineering, it is also an acceptance that *sustaining literature* is not something that we can choose to do or not do. What sustains itself, what is sustained, is the effect of forces of inscription that are also forces of destruction, erasure and occlusion. This may seem both abstract and irresponsible, but it is perhaps the only form of material responsibility possible. If, today, there is a 'we' who confronts Nature with a sense of loss it is because of an ongoing history of inscription, including all the technologies that have enabled human social assemblage, technology and literature in the narrow sense. Any 'we' who seeks to sustain itself does so by way of inscription. *Literary* sustainability is not some luxurious or privileged add-on that 'we' might choose to maintain, in addition to technology: everything that appears as a 'we,' as 'techne,' as engineering, as Nature, as past and as future is the outcome of inscriptive processes that select and erase, generating what will survive and be sustained.

In 'Biodegradables' Jacques Derrida (1989a) outlines a curious 'logic' (or counter-logic) of the relation between literature and sustainability. On the one hand, a genuinely timeless literary utterance would be biodegradable: one would write, inscribe and allow a sense to be incarnated in some material form but its meaning would be so monumental as to be assimilated into culture completely; even if all material instances of the utterance were to be destroyed, the event of meaning would have become so much a part of the culture as to no longer require the specificity of an archive. At its limit one might think of religions of the book as – supposedly – requiring the written version of the word only insofar as the full ideal of biodegradability has not been met. One needs written, remembered or stored versions of the word only because the word has not become fully interiorised or 'ingraven on hearts' (Milton 1953–82, ii.276). Or, as Wordsworth laments in the fifth book of *The Prelude,*

> Oh! why hath not the Mind
> Some element to stamp her image on
> In nature somewhat nearer to her own?
> Why, gifted with such powers to send abroad
> Her spirit, must it lodge in shrines so frail? (Wordsworth 1968: 68)

Contemplating the complete erasure of the archive, Wordsworth considers literature to be timeless content that is unfortunately stamped on frail matter. Here, the *matter* of the text is akin to an accidental interruption that might be easily erased, and what makes literary utterance significant is (for Wordsworth) its immortality quite distinct from the book:

> Oftentimes at least
> Me hath such strong entrancement overcome,
> When I have held a volume in my hand,
> Poor earthly casket of immortal verse,
> Shakespeare, or Milton, labourers divine! (Wordsworth 1968: 71)

In this respect literature is at once immortal (resisting erasure, and therefore non-biodegradable), and yet also utterly fragile (a mere casket). One might also think of the material inscription of pure truth and logic as being biodegradable: the truths of science may have required some inscription to become conscious and memorised, but once incarnated the ideal would be that logic, mathematics, geometry and formal systems would become part of universal human reason *and not* specific cultural documents. By contrast, literature is literature only with a certain non-biodegradability: unlike formal systems or even highly unremarkable texts that appear and then disappear, literature sustains itself *beyond meaning* as a mark or trace that is always in excess of, or remains above and beyond, any of the communicated and shared senses that it enables. We may like to think of Shakespeare and Wordsworth as part of a virtual archive – so inscribed in collective consciousness that no actual material copies need survive in order for the conception of the self in *Hamlet* or nature in *The Prelude* to be sustained. But once we give the matter some thought the notion of the biodegradable becomes an impossible limit that exposes something essentially unsustainable at the heart of literature and life. If life were to be – in its natural, sustainable and proper mode – fully biodegradable, then we might think of the living being as coming into existence and passing away without leaving a trace. Such a pure being would be so fully attuned to its ecology that it would do nothing more than contribute to the flourishing of the whole of which it is a part; it would neither pollute, nor disturb, nor scar. If that were so, then there would not only be no archive – no fossils, no interwoven ecologies, no instances of ongoing responsiveness – we might say that there would be no life, *if* living is defined as sustaining existence through time.

To maintain and sustain a living form requires resisting, however minimally, absolute biodegradability. One might say that to be is to pollute or to make a mark on one's *milieu* (Serres 2010). A certain model of nature – a being that might live without creating a mark, scar, loss or point of inertia and immobility – is analogous to a certain model of writing and meaning. Just as it is possible to imagine an eternal and pure nature, sustaining itself through time and enduring beyond human finitude, one might also imagine a writing so lucid, true and coherent that its initial textual form might decay and yet its sense would remain. Nature would not generate anything that would scar or disturb its ongoing, self-adjusting auto-poetic life. Writing would achieve a form of shared communicative transparency that would be global, inclusive and post-ideological. What we say and write would be so fully understood and true that the material medium of conveying sense might wither away. Both of these ideals of (natural and cultural) biodegradability would be challenged by thinking the transcendental unsustainability that marks what has come to be known as life. In order for us to say that something lives, it must sustain itself through time, but in so doing such a living being must *not* be completely attuned to its environment; it must not only resist falling back into some supposed pure immediacy of life, but must also possess a tendency or intensity of its 'own' that would mark it as 'a' life and therefore sustained. This minimal difference that is not life as such must be temporally finite, for if it were eternal and without limit, then it would not be 'a' life. It is one's physical temporal body that enables the passage through time and possibility, while incarnation also delimits possibility. The literary object is definitively *both* meaningful above and beyond its sustained incarnations (and continues to exist through time beyond its decaying materiality, passing into the sense of a culture), but is *also* tied to the singularity of a necessarily degrading matter. The literary text is therefore the extreme or limit case of the counter-logic of biodegradability; it must at once be sustained through a distinct matter that is not coterminous with life in general, and yet must also (in order to be recognised and circulated) take part in a cultural ecology:

> On the one hand, this thing is not a thing, not-as one ordinarily believes things to be-a natural thing: in fact 'biodegradable,' on the contrary, is generally said of an artificial product, most often an industrial product, whenever it lets itself be de-composed by microorganisms. On the other hand, the 'biodegradable' is hardly a thing since it remains a thing that does not remain, an essentially decomposable thing, destined to pass away, to lose its identity as a thing and to become again a non-thing ... Can one say, figuratively, that a 'publication' is bio-degradable and distinguish here the degrees of degradation, the rhythms, the laws, the aleatory factors, the detours and the disguises, the trans-mutations, the cycles of recycling? Can one transpose onto 'culture' the vocabulary of 'natural waste treatment' – recycling, ecosystems, and so on – along with the whole legislative apparatus that regulates the 'environment' in our societies? ... According to such a 'logic,' whose pertinence is, I believe, considerable but limited, nothing is destroyed and thus no 'document' 'biodegrades,' even if it is, according to some criterion or other, the most degraded or the most degrading. (Derrida 1989a: 813–14)

One of the ways in which we might think about writing, inscription and literature (especially after the advent of deconstruction) is as offering a form of radically futural, and possibly sublime, promise. Such a promise would have to do with forward mobility *and* sustainability. According to a Derridean conception of deconstruction, in order to live the present *as present* – as here, now for me – one must have already marked in the present what would be repeatable into a future. The present, by way of being lived, is already haunted by a past that it retains and a future it anticipates. One might refer to all experience, in its sustainability, as 'naturally' sublime; to experience the world as real and present (as sustained beyond the immediate 'now') one anticipates its existence beyond all the perceptions that I (and others) have of its phenomenal reality. If there is a world that appears then there is necessarily an accompanying sense of that which appears; one never grasps the world itself, even if it must be presupposed as the receding and withdrawing condition of experience. The real would be such that it always exceeded any inscribing mark or determination. One might say that the very constitution of the world as sustainable, or as remaining the same through time, generates a presupposed 'we' and a future that could not be constrained or determined by the very humanity that it implies: this is why Derrida, discussing Edmund Husserl's 'Origin of Geometry,' follows Husserl in arguing that the experience of the present presupposes a time and humanity beyond the present, even if that humanity cannot be reduced to any factual humanity and must always remain 'to come' (and even if that transcendental humanity always emerges from a specific inscription):

> But toward the end of the text, the Earth takes on a more formal sense. No longer is it a question of this Earth here (the primordial *here* whose factuality would finally be irreducible), but of a here and a ground *in general* for the determination of body-objects in general. For if I reached another planet by flying, and if, Husserl then said, I could perceive the earth as a body, I would have 'two Earths as ground-bodies.' 'But what does two Earths signify? Two pieces of a single Earth with one humanity.' From then on the unity of all humanity determines the unity of the ground as such. This unity of all humanity is correlative to the unity of the world as the infinite horizon of experience, and not to the unity of this earth here. The World, which is not the factuality of this historical world here, as Husserl often recalls, is the ground of grounds, the horizon of horizons, and it is to the World that the transcendental immutability attributed to the Earth returns, since the Earth then is only its factual index. Likewise – correlatively – humanity would then only be the facto-anthropological index of subjectivity and of intersubjectivity in general, starting from which every primordial *here* can appear on the foundation of the Living Present, the rest and absolute maintenance of the origin in which, by which, and for which all temporality and all motion appear. (Derrida 1989b: 84; author's italics)

Inscription marks the present as present and constitutes a 'here and now' that also surpasses itself, beyond human intentionality. There would be – in the very emergence of the here and now – *both* the constitution of a presupposed 'we' *and* the opening of that 'we' to 'any subject whatever', a humanity to come. Modes of inscription that mark the present *as present* cannot command or limit what might be carried over from the present to open new futures. Whatever a text or concept might mean or refer to in the present, it may always be reiterated into a future and open new horizons of possibility.

When Derrida argues for literature as the right to *say anything,* and for literature as a form of democracy, this (I would suggest) is because of a certain sustainability in the literary (Derrida 1992: 37).[1] In order for something *as sense* to be sustained through time, it must take on a material support that would liberate content from its immediate present: inscription allows the present to be carried over into a future, in the absence of the originating intent. It is by way of this law of iterability that Derrida criticises John Searle's notion of context as stabilising meaning: if a sign can operate in a context this is because it is repeatable and *not* reducible to the present (Derrida 1977). But the relations among sense, context, sustainability and biodegradability are themselves different across contexts, registers and epochs. As Bernard Stiegler (2011: 76) has argued, the Western tradition of *hypomnemata* does not simply allow the past to be retained into the future, but establishes an archive, and deploys technologies of retention that enable complex structures of sense to be sustained and developed through time. While the sense of mathematics intends a truth that would persist beyond its material inscription, and therefore aims for a maximal formalisation of inscription (where it is the impersonal operation that would be sustained over time), this is not so for the literary text. If, for example, every concrete and material instance of William Blake's engraved works were to be destroyed, the loss would be absolute and could not be retrieved (Derrida 1984). The literary text is tied to a proper name and materiality or signature that is never reducible to the concrete individual existing in historical time. 'Blake' or 'Shakespeare' designate complex inscriptive systems and modes of reading sustained through time, where one reads the signature as the sign of a once-present experience. The same is not true for mathematics, at least in its meaning. Imagine two scenarios: if the human 'archive' were destroyed one could still imagine mathematics and logic being 'discovered' or emerging again in their current form. By contrast, no poetic event could yield William Blake (with all the specific material supports of ink, copper, paper, wax and acid). While the mathematical or logical formula relies upon a subjectivity and truth in general that would remain the same through time, literature *sustains a marked difference.* If a community agrees that the geometry of Euclid is true, then the theorems may circulate and survive in the absence of Euclid, and well after the destruction of Euclid's text. A sense and ideality can be released from its origin, and even though it will require some inscription, it is not tied to a singular inscription. By contrast, literature is not only tied to the specificity of inscription, it is also undecidable as to how one might think the relation between sustainable sense and dispensable text. Perhaps nothing would be gained (in terms of meaning) if one were to discover an author's diary, notebook, shopping list or letters; but it is possible that some future reader might transform a text that we all 'know' by discovering one missing word, one mis-transcribed letter. We may all agree, for example, that Blake's 'Little Black Boy' is written with the sense or spirit of an ironisation of Enlightenment universalising humanism, and we might agree that *Hamlet* is about indecision, and *Heart of Darkness* is about colonialism, but the text remains and can always generate or promise other senses. Scientific sense aims to posit what would be true in the absence of any specific observer, while literary sense creates its unique and singular complexity tied to a mode of inscription that is at once material and yet never reducible to matter alone.

Perhaps no author forces us to confront the counter-logic of literary sustainability more than Blake. Rather than submit his poetry to the mass-producing printing industry, he engraved and painted every word of his prophecies, and thus allowed for the maximum preservation of his own intent, which would not be degraded by being formalised in a general sign system. And yet in order for this sense to survive through time, his poetry eventually had to be anthologised, and ultimately digitised. Preserving and sustaining Blake required that the biodegradable matter he had deployed would give way to the 'permanence' of the digital archive, even if the digital archive itself has a limited materiality *and* is very far from being biodegradable. To digitise a literary text is at once to recognise its material singularity – that it cannot just be memorised but must be stored – while also violating that material specificity. Digitising Blake disseminates and betrays his original intentionality; inscription is both preserver and destroyer.

What literary inscription generates is untimely repetition, or the power for voice to open to an unintended and unsaid future. Literature has a material and singular sustainability that is *suspended*; its ongoing existence through time allows for the release of contrary senses. A text may always be re-read, reframed, and reiterated without being anchored definitively in an intentionality or originating context. The literary would then present explicitly a potentiality for mobility and futurity that haunts all presence. To read a text as *literary* is to think of it as maximally sustainable, as what might always be reiterated beyond any of the present bounds of sense. A text would also be – as literary – hyper-promissory: we might 'know' (historically) that it makes no sense to read *Twelfth Night* as queer or transgendered (if the concept of queerness had no meaning in the original context), but the play allows for new performances that – say – most Americans would not grant to other texts such as the Declaration of Independence or the US Constitution, and that most Christians would not grant to the Bible. To read a text in a literary mode is to recognise a material sustainability that is destructive of any constituted sense; to repress or resist the literary is to posit a sense that sustains itself through time, beyond inscription. To read in a counter-literary manner would be to insist on the truth or sense of a text such as the Bible, the US Constitution, or Magna Carta, regardless of the language or rhetoric of its incarnation. To read in a literary mode is to focus on materiality rather than what Paul de Man referred to as 'phenomenalisation': one regards the inscription as the sign of something other than itself, as though it might be swept away to reveal something that would appear as its full sense (de Man 1996: 111).

In this respect, one mode of promissory deconstruction might be captured by Avital Ronell's conception of the telephone book: precisely in its detachment from a single commanding order, and its absence of address to any isolated individual, it is the most open of texts, enabling all forms of connection and solicitation (Ronell 1989: 5).[2] The telephone book is not *Hamlet* and therefore seems to be meaningless, random and void of sense, and yet this very emptiness also typifies a certain 'literary quality', liberated as the book is from any specified message. A telephone book is counter-literary (anonymous and random), but also hyper-literary – capable of addressing anyone, and capable of generating a call and message that appears as if destined for me. This open and futural dimension of a text offering almost anything while saying nothing might be a way of thinking about inscription as such. Too direct an address and/or delivery exhausts the message immediately; pure communication would be direct and intuitive. But if a text does not have a prescribed origin or destination, then it demands to be read. Even if, when one reads, the text appears to be speaking to me and me alone, this is only because the text was inscribed, set apart and rendered sustainable such that it might (later) arrive and offer itself to be read. A text that *speaks* requires (as Bernard Stiegler has argued) mystagogy: a belief that the sign is the mark of an authority whose spirit might be retrieved (Stiegler 2010: 26–38). All texts, especially when they seem to be prescient or timely, come from the future: if they call to me, offering sense, seeming to be rich with voice, then this is only because their immediacy seems to promise more than the absolutely punctual. Such a mode of thinking texts as futural would be sublime in at least two registers: it might intimate a sense of something far more deeply interfused of which the text is a broken-off fragment and which might always be unearthed. The text, in the manner of a nature whose law transcends our finite grasp, might offer a time beyond our grasp. To read a text in a promissory manner would allow us – after Derrida – to say wonderfully elevating things, such as 'literature is democracy'.[1] The text, because it sustains itself beyond the immediacy of voice, can always open to other voices.

The first mode of sublimity would lie in the text's capacity – whatever the present – to *open an infinite.* It is because texts must submit to systems of inscription that are *not those of the self-present immediacy of the living voice of the author* that they can be re-read, beyond any already given intention or experience. Another, second, mode of sublimity would insist less upon an intimated openness and expansiveness of the text, and more upon a materiality that would be resistant to our anthropocentric imaginings. (This would be Paul de Man's material sublime (Warminski 2001).)

We could think of materiality (after Derrida) as that which is iterable: to submit to a system of traces is to tear sense from itself, and open it to a future not its own. In this case matter is what enables a certain taking on of form – such as Blake's words being engraved, with voice articulating itself through ink, paper, copper, wax and acid. The voice is not just incarnated in matter; the form of matter generates a style of repeatability. Blake's engraved words can be anthologised, memorised, translated, coupled with other texts (biography, history, medical theses, prison manuals, maps). The matter of inscription always exceeds its present, and even the absolutely singular text – the engraved pages of Blake residing in the New York Public Library – takes its form from a system of differences that allows the text in its isolation to be more than itself. Blake engraved every word of his work, but he still used the formal system of English (despite his many invented words). In order for the matter he worked upon to be formed and make sense, he drew upon another materiality, the instituted system of differences of the English language and the phonetic alphabet. We might say (following Deleuze and Guattari) that rather than think of matter that takes on and enables a surviving form, there are formed matters – with forms being the way in which matter becomes a substance, and with matter being the way a form can come into being:

> He used the term *matter* for the plane of consistency or Body without Organs, in other words, the unformed, unorganized, nonstratified, or destratified body and all its flows: subatomic and submolecular particles, pure intensities, prevital and prephysical free singularities. He used the term *content* for formed matters, which would now have to be considered from two points of view: substance, insofar as these matters are 'chosen,' and form, insofar as they are chosen in a certain order (substance and form of content). He used the term *expression* for functional structures, which would also have to be considered from two points of view: the organization of their own specific form, and substances insofar as they form compounds (form and content of expression). A stratum always has a dimension of the expressible or of expression serving as the basis for a relative invariance; for example, nucleic sequences are inseparable from a relatively invariant expression by means of which they determine the compounds, organs, and functions of the organism. To express is always to sing the glory of God. (Deleuze and Guattari 1987: 43, authors' italics)

There is a substance of form and a substance of content, and a form of expression and a form of content: what one says is enabled by certain forms that can be articulated in matters, while matter gives that formed content the possibility of being repeated in other matters. Matter would be mobile and would be given *as matter* only by way of differentiating itself, and would allow difference and articulation to appear only by way of its material distribution. The sublimity would lie in the capacity for the forms inscribed in matter to be released from the matter in which the form first appeared, *and* for the matter through which form appeared to be de-formed. This is how Derrida reads both Joyce and Husserl, aware that literature in its Joycean mode aims for maximal inclusion of all the forces of the material *word*, inscribing all the potentialities of the sign in a single book, while phenomenology would aim to intuit a sense beyond all the singular incarnations that would be present for any subject whatever.

> Since equivocity always evidences a certain depth of development and concealment of a past, and when one wishes to assume and interiorize

> the memory of a culture in a kind of recollection (*Erinnerung*) in the Hegelian sense, one has, facing this equivocity, the choice of two endeavors. One would resemble that of James Joyce: to repeat and take responsibility for all equivocation itself, utilizing a language that could equalize the greatest possible synchrony with the greatest potential for buried, accumulated, and interwoven intentions within each linguistic atom, each vocable, each word, each simple proposition, in all worldly cultures and their most ingenious forms (mythology, religion, sciences, arts, literature, politics, philosophy, and so forth) …
>
> The other endeavor is Husserl's: to reduce or impoverish empirical language methodically to the point where its univocal and translatable elements are actually transparent, in order to reach back and grasp again at its pure source a historicity or traditionality that no de facto historical totality will yield of itself. (Derrida 1977: 102–3)

Just as Kant's sublime seems to be as much about a Nature that surpasses all attempts to describe it as it is about an immateriality of thought that can anticipate an infinite beyond given nature, so Derridean inscription at once promises a future released from inscription – a 'justice to come' – only because inscription will allow the word of justice to be repeated beyond any of its current senses. However, this conception of inscription and materiality *cuts both ways*. The inscribed word may be read as a fragment cut off from a future of reading that one cannot contain; one may read in order to discern the spirit of the origin, but one might also imagine the same text existing in the future. Insofar as it is inscribed, the word is necessarily capable of being cut from its origin. The supposed 'origin' that we read in any text is already cut apart from itself. It may or may not open to future readings; it may promise openings to new sense, but it may also persist into a future without readers (without humans, without a people).

This double-sidedness of inscription allows us to consider another sublime or another sustainability: not a sustainability of sense but a non-biodegradable mark. Yes, one can read Blake or Shakespeare through a framing *milieu* of postcolonialism, or open the text to any number of scales – the history of race, sex, objects, bodies – but one might also *refuse* to grant the text such futural mobility. Here, we might think of the two modes of sublime as discussed by Paul de Man: a natural, and then a technical or poetic sublime (de Man 1996: 70; Newmark 2012: 95; Warminski 2013: 47–9). The former appears as if nature were so overwhelming as to signal the limits of present inscription, thereby allowing the power *to inscribe* to reach beyond itself, *as if* it might think beyond its inscriptive present. But there is also a technical and non-recuperative sublime, where the power lies in inscription's erasure of itself, using some matters to displace others. As de Man describes it (following Neil Hertz's reading of Longinus) the rhetorician creates some figures so powerful

– so illuminating – that they bathe the scene with light and conceal rhetoric and figuration. The example in Longinus is Demosthenes, who hails the vanquished in the style that might be appropriate for a funeral oration, and it is the style that then covers over *what is being addressed*. The marks and styles operate to produce a tone of victory despite defeat and are sublime because they erase their own working. What is sublime is not the content that exceeds inscription but inscription's configuration and displacement of its own matters: 'Surrounded by brilliance, the lesser light (the artifice) disappears from sight. But what is it that disappears? Not the particular figure, of course, but "the fact that it is a figure" – its figurativeness, so to speak' (Hertz 1985: 17). One could describe this manoeuvre, as Neil Hertz does, as a light so blinding that it drowns out all lesser lights. But in doing so – describing figuration through the figure of light (that is itself inscribed materially) – a figure stands in for and explains all other figures. One could say that what occurs is a certain violence where marking erases its own operations by producing technical figures that appear as something 'like' light, a *milieu* or medium of seeing, a mobilisation or illumination. Here I would like to think about de Man's *material sublime* which is the immobilisation of all these events of displacement and figuration, where all one has is matter and matters 'side by side'.

What if one were not to think of matter as the incarnation of a form that could be repeated in other matters? What if inscription were to occur once and for all, and be nothing more than the formed matter that it simply is? A text might operate less by intimation – less like a nature whose depths exceeded any single vision, and less like a living organic matter that always harboured the seeds of becoming and futurity – and more by way of theft. By theft I refer both to the severing of a future – *this* is inscribed and not something else – and also to theft as a form of trickery or 'brigandry' where a figure stands in for and commands figuration. What if cutting, truncating and eliding were ways of ensuring that a text would not be examined *too* closely? It is only by disabling the theoretical gaze that one could read the text as if it harboured a sense and future. One might say that to read a text – *any* text – as *necessarily* promising new futures and unimagined horizons relies on deflecting and erasing the text's scars. To read and find sense is to exclude and edit, with elisions ranging from typographical errors to minor, peripheral or nonsensical marks. It is to see the ink not as ink. One might not find Shakespeare too readable, and one might not have a timeless Shakespeare, if one were to spend too much time on 'his nose was sharp as a pen and a babbled of green fields' (*Henry V*, ii.3). Rather than see literature that allows one to say anything, one might see literature as that which – in *having being inscribed* – gives us only matters to be read.

This opens two senses of sustainability and promise: a text is sustained into the future and promises infinite futures if its matter is such that it releases other material iterations. But a text is also sustained in a certain non-biodegradability; a text is *this inscription, and no other, and may promise a future that is not what we want it to be.*

Here, we might think of the Anthropocene. When the concept was first formed it seemed to promise so much for the humanities, as a mode of inscription beyond the hand, beyond intentionality, *and* beyond the containment of inscription in human scales. On one understanding of the sublime, one might think that this natural object – an inscription that intimates a movement of history that is not of our own hand and that we cannot erase – as both a conditional promise and a violent cut. *If* 'we' can read certain traces as the marks of what we will have been, then we might say that we are opened to a power of tracing beyond humans, who are no longer the sole authors of sense. A history and a future *has been written*, and this passive voice might open something like the Derridean 'perhaps.' No trace or mark is in command of itself; nor can any inscription or context determine in advance what other marks it might enable. This rogue power of inscription opens another de Manian counter-promise. One might – as I would suggest we ought to do – tie the history of sense (from Euclidean geometry to modern techno-science) to a collateral inscription from which 'we' *cannot* be detached. Inscription is non-biodegradable: the very techno-science *and* liberal humanist archive that allows us to read the Anthropocene is also bound up with what the Anthropocene promises: there may well be no future, and there can certainly be no future for the private reading subject whose power has been generated from an earth that now is indelibly inscribed. The radical promise of the Anthropocene – that it might open inscription beyond intentionality and the hand, and that its sublimity might not be that of human thought recuperating itself through the finitude of inscription – has not been fulfilled. On the contrary, the Anthropocene has operated as something like the technical or rhetorical sublime that allows a figure to be so blindingly illuminative as to erase all other figurations. It has become an inscription of hyper-phenomenal proportions that has cut off reading. The geological layer is frequently *not read* as trace, inscription or indelible mark but appears as an imperative, telling us who we are and what we must do.

To anticipate what follows, one might think of all the ways in which utter defeat and loss (in the Anthropocene epoch) have been presented as sublimely human and futural. The current predicament of climate change is – for Naomi Klein and for less ethically motivated corporations – an opportunity for a new future (Klein 2014). The new genre of cli-fi repeatedly presents a wasted planet and defeated humanity as precisely the *milieu* and backdrop for an elevated sense of human sublimity (where the sublime renders a ruined present into an imagined glorious defeat). Both Chris Nolan's *Interstellar* (2014) and Neill Blomkamp's *Elysium* (2013) begin with a world defeated by anthropogenic climate change, and with a humanity vanquished by climate-opportunist corporations. The substitutions, tricks and thefts of Anthropocene inscription that occur with the short-circuiting of reading are not only evidenced in corporate dreams of geoengineering and other declared states of emergency that allow the geological scale to trump and blind all other inscriptions, but also resonate in many responses of the humanities that will have their day in the sun now that the Anthropocene draws us back to the common fate of a humanity to come.

Just as contemporary cinema is increasingly figuring planetary catastrophe in military terms, and then shifting attention from *what* is being destroyed (nature, earth, life) towards an act of heroic sacrifice, so the Anthropocene has reinvigorated theory, allowing yet one more scale or register that permits inscription in its literary sense to survive into its own future. Apostrophes or truncations that manufacture the grandeur of those vanquished – eliding the negativity of loss – are the condition for a broader rhetorical industry of blindness. If there are science-fiction narratives that present a benevolent (defeated) humanity ultimately triumphing over those who plundered and destroyed the planet (or who seized the spoils after planetary destruction), then such fantasies occur amidst presentations of 'our' supposedly doomed future as, nevertheless, calling upon 'us' to become sustainable. From geoengineering, to claims regarding the politically revolutionary opportunities enabled by climate change, calls that ask us to act now rely on a refusal of reading, and rely on seeing the geological register of the Anthropocene as *immediately political.*

As Gilles Deleuze and Félix Guattari (after Nietzsche) argued, the way in which memory becomes moral, or the way in which 'desire' generates a human promissory animal – sustaining one's commitments into the future (by owing, paying, valuing, being able to say 'I') – is by inscription: 'The prime function incumbent upon the socius, has always been to codify the flows of desire, to inscribe them, to record them, to see to it that no flow exists that is not properly dammed up, channeled, regulated' (Deleuze and Guattari 1977: 48). It is not relations that are relayed by inscription, but inscription that generates relations. One might think that there is a huge distance separating the literary/textual emphasis of deconstruction – particularly in its de Manian mode – from the vital realism of Deleuze and Guattari, but I would suggest that de Man's materialism should prompt us to think otherwise. What de Man refers to as 'phenomenalisation', or seeing inscription as bearing a relation to a prior ground of which it is the sign, might be compared to what Deleuze and Guattari refer to as the despotism of the signifier, or the notion that inscription is the expression of a preceding whole that it mediates (even if that whole is ever receding). Against this, they argue for a general inscription from which relations and territories are generated such that *it is the mark that makes the territory*:

> The territory is not primary in relation to the qualitative mark; it is the mark that makes the territory. Functions in a territory are not primary; they presuppose a territory-producing expressiveness. In this sense, the territory, and the functions performed within it, are products of territorialization. Territorialization is an act of rhythm that has become expressive, or of milieu components that have become qualitative. The marking of a territory is dimensional, but it is not a meter, it is a rhythm. It retains the most general characteristic of rhythm, which is to be inscribed on a different plane than that of its actions. (Deleuze and Guattari 1994: 315)

If we think of Deleuze and Guattari as materialists, theirs is not a vital materialism such that *life* can be posited as the ground from which signification emerges, for life itself is effected from 'rhythm' or the generation of distances. Matter is generated from manners. What *is* – ontology – emerges from relations, with relations being extrinsic; the relations that a being will enter into depend on encounters, and are not intrinsic to the being in question. A being is nothing more than its ongoing and dynamic encounters, even if there are tendencies not exhausted in those encounters or relations. Theirs is not a world at a distance from what is perceived, but a world that is all the perceptions that generate spaces and times. Deleuze and Guattari's commitment to the exteriority of relations precludes any term or matter from being the origin from which relations emerge; rather, it is inscription or the marking out of distances that generates matters. What they say about the book, then, is part of a more general claim about rhythm *from which* terms or matters are generated:

> A book has neither object nor subject; it is made of variously formed matters, and very different dates and speeds. To attribute the book to a subject is to overlook this working of matters, and the exteriority of their relations. It is to fabricate a beneficent God to explain geological movements. In a book, as in all things, there are lines of articulation or segmentarity, strata and territories; but also lines of flight, movements of deterritorialization and destratification. (Deleuze and Guattari 1987: 3)

Here, we can draw upon Warminski's account of de Man and the sublime. What we know and see *as matter*, as the things of this world (including light bulbs, paint, television screens, projectors, rays of sun) can always be presented as a figure for the ways in which matter comes to appear; 'a' matter comes to stand for materiality in general. The dispersed plane of inscription is explained from one of its effects: we might think of the light of the sun as that which allows matter to appear as matter, but what occurs in this substitution is not just the use of one material object to explain material relations and appearance, but a concealment of all the other 'matters' (paint, screen, ink) by which this figure of the sun is separated and presented. According to Warminski:

> For when the relation between the sublime and figure is revealed to be not like the relation between the greater light and the lesser light – not like the relation between phusis and phusis, and not even like the relation between phusis and techne – but rather like the relation between paint and paint, then it becomes legible that the relation between the sublime and figure is a relation between figure and figure, i.e., techne and techne. (Warminski 2014)

One might say that rather than a nature that unfolds from itself and that can be figured by techne, with the sublime signalling the irreducible gap between nature and figure, there are just gaps and distances that enable *appearance*, if appearance is understood as the appearance *of* something other than the play of matters. Warminski then concludes with a claim for *reading* over phenomenalisation: what we have – what appears – are matters, but the ordering of some matters as standing for or, appearing to be, figures of other matters (with some figures explaining other figures, and some figures appearing as so bright as to be invisible) occurs in an event of reading which, 'ideally', might appear as an eye that would only see what is present at hand, and not recuperate matters as substitutions of lost spirit.

The notion of a natural sublime – a power in nature that is so overwhelming as to threaten dismemberment but that can be recuperated by representing nature's power *as unsurpassable* – occurs by means of a series of technical and material events. One of the ways the Anthropocene has been figured is as a natural sublime, an inscription of a time and power of figuration beyond the human that can, nevertheless, be recuperated by the very event of humans reading their own finitude. This recuperation can take many forms and registers.

First, one might argue that the inscription of the Anthropocene gives the lie to the modern enclosure of the knowable within the human given; now there are perceptible signs of what occurs beyond perception and synthesis. However, in this new literalism – declared variously in terms of new materialisms, new vitalisms, or material and vital turns – what is cut out or erased (biodegraded) is something like techne. Matter is given as that from which inscription emerges, but this conception of a pre-inscriptive materiality may yield two modes, or two promissory gestures. One might – as I will hope to do – think of materiality as that which promises in a mode that is not only threatening, but also metaleptic: the present harbours a potential to be sustained into the future in a manner that is at odds with the apparent present it supposedly extends. What is now being read phenomenally as the Anthropocene – as a record of a past that promises an inhuman future – seems to open *both* a sense of matter as a sign of that which calls for, generates and confirms human agency *and* as an indication of an agency-without-agency that persists or is sustained in indelible inscription.

Why are some matters privileged as the sign or appearance that allows us to read all other matters; how have we stopped *reading* by seeing some traces as ways of closing down counter-inscriptions? Against a matter that can be read as offering a narrative of human sustainability, one might think of matter as radically counter to any forms of biodegradability: matter (regardless of how we might discern its human promises) sustains itself beyond all our thoughts of a world that would be in accord with our imaginings. I would therefore – after de Man – challenge the epistemology that has accompanied the Anthropocene – the notion that some traces or stratifications close down questions simply by presenting themselves as that which can be known. Does the Anthropocene *really* give the lie to the multiple tracings of the world? Does it *really* demand that we think of humans as a single species, and as a species who must act now and in concert? Rather than offering itself as a knowledge that trumps all speculation and confirms what we ought to do to sustain ourselves (where the Anthropocene would be the sign that offers the frame for reading all other signs), the Anthropocene might cause us to think of inscription as a materiality that is unreadable. Whereas some conceptions of the Anthropocene regard the geological register as an inscription that eliminates all previous vagaries regarding the sense of the earth and therefore erases the need to *read*, one might say that such definitive knowledge is only possible by way of concealing all the interpretive manoeuvres that allow one stratum to become the privileged frame for all others.

Second, one might say that the very notion of epistemology is enabled by a figure of the play of lights: in traditional Platonic and Enlightenment conceptions of knowledge, rather than be captivated by the shadows cast by light, one should turn the soul around to the condition of the visible. (Here, one might ask about the conditions for the materiality of such a scene: who paints the tableau of privileged and lesser lights, and who displays the scene of proper seeing?) This is a question asked by theorists as diverse as Luce Irigaray and Bernard Stiegler: how is the frame through which the world appears as already fully readable inscribed and sustained? The archive is not only a body of texts to be read, but can appear as an archive only by way of an established scene and frame of questioning. Stiegler (2013) refers to an 'arche-cinema' that is also a mystagogy, a belief that there is a sense to be read and sustained and to which one submits. Irigaray (1985) famously regards this positing of a matter to be read by a subject as the very structure of sexual sameness: there is a world that has been set apart from me and that can be grasped and known by way of its appearance to me. De Man talks about the *defacement* that occurs with the production of an ongoing and readable sign and, writing through Rousseau, argues that, 'Behind the stability and the decorum of private law lurk the "brigands" and the "pirates" ... whose acts shape the realities of politics between nations, the most difficult adjustment being the necessity of considering these mixed standards as entirely honorable' (de Man 1976: 665). The common space of the law does not follow seamlessly from 'nature'; nature is effected as that which generated the law only after the inscription of the law.

Before asking questions about the legitimacy of this or that law, and before negotiating how 'we' might respond to threats to 'our' sustainability, 'we' need to think about the genesis of the 'we' or of the ways in which what de Man refers to as 'phenomenalisation' emerges: what allows the world to appear *as* the sign of some available knowledge that would, in turn, enable a domain of expertise and managerialism? A similar question was asked by Foucault (1970: xix) in *The Order of Things* regarding the 'table' or plane across which the knowledge of objects is distributed. Foucault argues that one might think about language's own 'shining' or the way in which inside and outside, visible and articulable, are differentiated to produce a structure of knowledge. To think about the 'shining of language' as that which precedes the plane or distribution of knowledge suggests that perhaps one might think of language as akin to a light that distributes the sensible and intelligible (Foucault 1970: 369). Or perhaps we might think of light as akin to a metaphor that enables distinctions between visible and hidden. When de Man writes about the sublime he shifts attention from a nature whose might can only be grasped by way of the play of figures, to a play of figures that generates a privileged figure – such as light – that appears as the natural medium of all appearing. If one questions the ways in which geological stratification has come to appear as the tracing that discloses the logic of all other tracings, then we arrive at two possibilities: either 'we' accept that we become truly aware of the structure of 'man' and his relation to the world (finally) with the illuminating appearance of geological framing, *or* we might read the scene of the Anthropocene. Reading, here, is not the seamless passage from inscription to knowledge, but the confrontation with inscription 'as such', even if the 'as such' is already a recuperation of inscription's distance and difficulty.

In a quite distinct way, then, the twilight of the Anthropocene – or the appearance of the destructive structure of appearance – prompts the question of the distribution of the sensible, or how it is that some lights appear as signs of what is to be done. What has enabled what has come to be known as the Anthropocene is a recuperation, constantly, of inscription. It is as though the play of lights, matters, distances and figures were there to yield the truth of the world. What has been occluded is the technical and rhetorical series of ruses and substitutions that allows one play of figures to establish itself as the figure of all appearing. There must have been, or will have been, a relation among matters in order for a light to appear as the source that enables figuration to be seen.

As humanity begins to witness its possible end and orients itself towards an inscription beyond its own hand, it might begin to see the journey of enlightenment and of turning the soul away from captivation by shadows as enabled by a material support that becomes visible only in its disappearing. Far from being the ground towards which all knowledge ought finally to tend, one might think of the light of reason as the trope that occluded all other figurations. Very crudely, if practices of reflection, critique, and technology in the narrow sense, rely on a harnessing of matters from elsewhere, such dependence and secondariness can only be known in the moment of its loss. The industries and institutions of critique (including universities, the bourgeois public sphere, media, publishing, dissemination and the habits of private reading) have been possible only because of a history of energy extraction: enlightenment requires a prior history of resource seizure, beginning (at least) with the Athenian polity's dependence upon slave labour. Today, when those energy resources and practices of seizure appear to be no longer sustainable, we arrive at the illumination of the Anthropocene: the grand project of industry, reason and progress with all its metaphors, figures and lights is bound inextricably to unsustainability.

And yet – finally – it is by addressing loss, or by apostrophising this loss, that a certain technical sleight of hand generates one final sublimity. These rhetorical manoeuvres might range from cinematic cli-fi epics, such as Chris Nolan's *Interstellar*, to the current condition of what I will refer to as theory refuge. At the moment of the impending loss of life the triumph of definitive knowledge appears. The Anthropocene as an epistemological ruse has generated all forms of knowledge managerialism and priesthood. In the meantime, cultural production increasingly focuses on the literal end of man to stage one final call to arms. In both cases one departs from a defeated terrain, and yet it is the very capacity to mourn or apostrophise loss that transfers power to the event of losing (where losing is ambiguously poised between the sense of something that has been lost, and the loss itself as the appearance of the truth of all appearing). The 'end of the world' functions as our way of finally knowing the truth of our own destructive history, while dramatisations of the 'end of the world' draw heavily upon figures of good versus evil, and a final affirmation of heroic triumph. If the Anthropocene is 'our' twilight, then this is both because the sense of life appears only in the moment of the loss of the light of life, and because a privilege accorded to one stratum of inscription dims all other lights.

As what thinks of itself as the species begins to sense that it is in a stage of twilight and begins to consider its own end within time, we might begin to ask what forms of viewing, visibility and illumination have enabled species-reflection. That is, one might think of the Anthropocene in the mode of what de Man referred to as 'phenomenalisation', as a sign that offers – finally – true knowledge of the world as it is, and – in turn – explains the emergence of all other forms of inscription. (In this respect the geological scale of the Anthropocene would frame the emergence of life, cognition, humanity and its self-reflective triumph). Alternatively, and preferably, one might *read* the Anthropocene: there would be no direct passage from inscription to knowledge, *nor* to a humanity that would be the revealed ground or 'we' to whom the signs of the earth would be addressed.

Notes

1 Literature is democracy for Derrida because the literary text is manifestly tied to the singularity of its inscription. We can paraphrase, translate and summarise William Shakespeare's *Hamlet* or Zadie Smith's *White Teeth,* but the text remains with a material resistance that does *not* mark mathematical or logical texts, which (ideally) have a truth regardless of the language or text used to convey their sense. Literature can say anything because its textuality is not bound or constrained by intentionality or context. Because the text itself – and not meaning, intention or truth – is what survives, literature can always be re-read, can be opened in other contexts and can deploy a plurality of voices not attributed to the author's intentionality. See Culler (2008).

2 To the extent that you have become what you are, namely, in part, an automatic answering machine, it becomes necessary for questions to be asked on the order of: Who answers the call of the telephone, the call of duty, or accounts for the taxes it appears to impose? Its reception determines its *Geschick,* its destinal arrangement, affirming that a call has taken place. But it is precisely at the moment of connection, prior to any proper signification or articulation of content, that one wonders, Who's there?

References

Culler, Jonathan 2008. 'The Most Interesting Thing in the World', *Diacritics* 38 (1–2) (Spring–Summer 2008): 7–16.

De Man, Paul 1976. 'Political Allegory in Rousseau', *Critical Inquiry* 2 (4): 649–75.

De Man, Paul 1996. *Aesthetic Ideology*. Ed. Andrzej Warminski. Minneapolis: University of Minnesota Press.

Deleuze, Gilles and Félix Guattari 1977. *Anti-Oedipus: Capitalism and Schizophrenia*. Trans. Robert Hurley, Mark Seem and Helen R. Lane. New York: Viking Press.

Deleuze, Gilles and Félix Guattari 1987. *A Thousand Plateaus: Capitalism and Schizophrenia.* Trans. Brian Massumi. Minneapolis: University of Minnesota Press.

Deleuze, Gilles and Félix Guattari 1994. *What is Philosophy?* Trans. Hugh Tomlinson and Graham Burchell. New York: Columbia University Press.

Derrida, Jacques 1977. *Limited Inc.: Abc.* Baltimore: Johns Hopkins University Press.

Derrida, Jacques 1984. 'No Apocalypse, Not Now (Full Speed Ahead, Seven Missiles, Seven Missives)'. Trans. Catherine Porter and Philip Lewis, *Diacritics* 14 (2): 20–31.

Derrida, Jacques 1989a. 'Biodegradables: Seven Diary Fragments'. Trans. Peggy Kamuf, *Critical Inquiry* 15 (Summer): 812–73.

Derrida, Jacques 1989b. *Edmund Husserl's 'Origin of Geometry': An Introduction.* Trans. John P. Leavey Jr. Lincoln: University of Nebraska Press.

Derrida, Jacques 1992. *Acts of Literature.* Ed. Derek Attridge. London: Routledge.

Foucault, Michel 1970. *The Order of Things: An Archaeology of the Human Sciences.* London: Tavistock Publications.

Hertz, Neil 1985. *The End of the Line: Essays on Psychoanalysis and the Sublime.* New York: Columbia University Press.

Irigaray, Luce 1985. *Speculum of the Other Woman.* Trans. Gillian C. Gill. Ithaca: Cornell University Press.

Klein, Naomi 2014. *This Changes Everything: Capitalism vs. the Climate.* New York: Simon & Schuster.

Latour, Bruno 2011. 'Waiting for Gaia: Composing the Common World through Arts and Politics'. Lecture at the French Institute, London. November. www.bruno-latour.fr/sites/default/files/124-GAIA-LONDON-SPEAP_0.pdf. Accessed 15 February 2017.

Milton, John 1953–82. *Complete Prose Works*, Vol. 5. Ed. Don. M. Wolfe. New Haven: Yale University Press.

Newmark, Kevin 2012. 'Inscribing the Political'. In *The Political Archive of Paul de Man: Property, Sovereignty and the Theotropic.* Ed. Martin McQuillan. Edinburgh: Edinburgh University Press, 91–102.

Ronell, Avital 1989. *The Telephone Book: Technology–Schizophrenia–Electric Speech* Lincoln: University of Nebraska Press.

Serres, Michel 2010. *Malfeasance.* Trans. Anne-Marie Feenberg-Dibon. Stanford, CA: Stanford University Press.

Stiegler, Bernard 2010. *Taking Care of Youth and the Generations.* Trans. Stephen Barker. Stanford, CA: Stanford University Press.

Stiegler, Bernard 2011. *The Decadence of Industrial Democracies.* Trans. Daniel Ross. Stanford, CA: Stanford University Press.

Stiegler, Bernard 2013. 'The Organology of Dreams and Arche-Cinema'. Trans. Daniel Ross. *Screening the Past*, June. www.screeningthepast.com/2013/06/the-organology-of-dreams-and-arche-cinema/. Accessed 15 February 2017.

Warminski, Andrzej 2001. '"As the Poets Do It": On the Material Sublime'. In *Material Events: Paul de Man and the Afterlife of Theory.* Ed. Tom Cohen. Minneapolis: University of Minnesota Press, 3–31.

Warminski, Andrzej 2013. *Ideology, Rhetoric, Aesthetics: For De Man.* Edinburgh: Edinburgh University Press.

Warminski, Andrzej 2014. 'Longinian Sublimes'. Unpublished manuscript.

Wordsworth, William 1936. 'Lines Composed a Few Miles Above Tintern Abbey.' In *Poetical Works*. Ed. Ernst de Selincourt. Oxford: Oxford University Press, 1936.

Wordsworth, William 1968. *The Prelude: Or, Growth of a Poet's Mind (Text of 1805)*. Ed. Ernest De Selincourt. Oxford: Oxford University Press.

Wordsworth, William 1985. *The Pedlar, Tintern Abbey, the Two-Part Prelude*. Ed. Jonathan Wordsworth. Cambridge: Cambridge University Press.

Part II

Reading sustainability

7

Collapse, resilience, stability and sustainability in Margaret Atwood's MaddAddam trilogy

Dana Phillips

> Myths are the agents of stability, fictions the agents of change. Myths call for absolute, fictions for conditional assent. Myths make sense in terms of a lost order of time ... fictions, if successful, make sense of the here and now. (Kermode 1967: 39)

> The good news is that the end is in sight. The bad news is that it's not happy. The worse news is that it's also not the end. (*@neinquarterly*, 5 February 2015)

Collapse, resilience, stability and sustainability: some definitions

With the publication of *MaddAddam* in 2013, the story that Margaret Atwood began with *Oryx and Crake* in 2003 and continued with *The Year of the Flood* in 2009 stands complete and can be read as a single narrative. Over the course of the MaddAddam trilogy, Atwood relies on notions of collapse, resilience, stability and sustainability as she establishes setting, navigates turns of plot and weighs the actions of characters, among which we must number the multinational corporations – OrganInc, HelthWyzer, AnooYoo and others – where several of her protagonists are employed. The concepts of collapse, resilience, stability and sustainability are also central to recent work in ecological theory and environmental history, where they emerge as crucial to the intelligibility of environment as such. Even, or perhaps especially, when they are not articulated explicitly – as is the case in Atwood's trilogy – these four concepts may serve as tropes, lending weight to and imposing structure on environmental narratives. So, given that Atwood does not acknowledge their importance to her narrative openly, and in light of the fact that 'the popular application of the information generated by ecologists is generally bad news' (Gunderson and Allen 2010: xiii), it is necessary for me to begin by defining collapse, resilience, stability and sustainability as carefully as I can.

For environmental historians, a collapse entails a drastic reduction in natural, cultural and social complexity, and is marked by a rapid drawdown of natural resources, a failure of crops, a disappearance of centralised government and an end to all public works. Some environmental historians portray collapse as the most dire possibility posed by environmental crisis, albeit one that can be offset by ecological stability and mitigated by natural and cultural resilience, if not circumvented altogether by the utopian possibilities of sustainability. Hence one might be tempted to argue, if one were an environmental historian, that collapse and sustainability are in something like a dialectical relationship, with stability and resilience serving as mediating terms.

That temptation should be resisted, since each of these four terms marks something of a theoretical and factual conundrum, if not a vacuum. Despite all the attention the notion has received, an actual collapse of the most drastic sort, one that gives rise to apocalyptic rumblings and post-apocalyptic recriminations, has never been documented, at least not in a wholly satisfactory fashion. The idea that such an event has occurred numerous times throughout human history, and might present a clear and present danger today, was popularised by Jared Diamond's 2005 book *Collapse*. Academically rigorous treatments of the subject make Diamond's account seem suspect on both factual and theoretical grounds (see Tainter 1988 and the essays collected in the 2010 volume edited by McAnany and Yoffee). Of course, that a total collapse of the kind Diamond writes about may *never* have occurred does not mean that no collapse of any kind can *ever* occur. Ecologists recognise that natural systems can only be subjected to so much stress before they collapse, but their use of the term is less apocalyptic and more precise than Diamond's. Here, for instance, is a description of what an ecologist might regard as a garden-variety collapse: 'Changes in either driving or state variables may cause collapses. Often, the system provides no warning, and collapse follows an unexpected, but often inevitable event' (Gunderson et al. 2010: 4). For an ecologist, collapse is simply something that happens to natural systems from time to time. They crash like a desktop computer, but they don't burn like the *Hindenburg*. Population dynamics afford a classic illustration of how collapse-like phenomena can be regarded as fundamental features of natural history: some species of fish, like the North Atlantic cod and the striped bass, endure irregular cycles of boom and bust, of random waxing and waning in their numbers. Managing such species means riding a wildly fluctuating logistical seesaw.

Sustainability is an even more ambiguous concept than collapse, though with sustainability the problem is more prospective than retrospective, as the discourse on the subject has been cast, for the most part, in the optative mood. 'Sustainability' has been a word in search of a precise meaning for several decades. Many people and most governments are convinced that sustainability would be a good thing. It is variously defined as an embrace of alternative energy sources, as the practice of so-called permaculture, as a philosophy of low-impact living, and – for some of its adherents, including most governments – as a new strategy for continued development along the familiar lines established by capitalism and in the wake of peak oil. Precisely because it means so much, no one has been able to disambiguate the term, to identify the necessary components of sustainability and explain how it might be achieved. As Hannes Bergthaller has observed – succinctly and pertinently, in an article on *Oryx and Crake* and *The Year of the Flood* – 'sustainability is a notoriously fuzzy term' (2010: 730; also see Caradonna 2014 and Grober 2012). I would suggest that its fuzziness is in direct proportion to the amount of time environmentalists – and entrepreneurs – have spent trying to define 'sustainability': usage has had the consequence, unintended but not unusual, of broadening the term's meanings.

As a concept, stability is something of a throwback to the ecology of yesteryear. Some natural systems – fewer than you might think – display their stable qualities when they emerge from a disturbance, if not from a collapse, and reestablish themselves in good working order (or return to an appearance, at least, of equilibrium) within a relatively short period of time. I say 'relatively short' because the period can range from a few days or weeks or months, as in the case of a riverine system returning to normal flows and population levels of its resident species after a flood, to years, as in the case of a clear-cut temperate forest returning to a state of mature growth and something closely approximating its original diversity of plant and animal life. It helps the process of recovery if engineers have not dammed and channelised the river and loggers have not clear-cut the forest, burned the slash and turned the altered landscape into a plantation for the production of lumber or pulpwood.

As a phenomenon, resilience is more complicated – and, it now appears, more common or, if you like, more *real* – than stability. A relatively new concept, resilience challenges the tidy assumptions that once led to a widespread belief in stability as a fundamental attribute of pristine natural systems. As developed by theoretical ecologists, the concept is somewhat counter-intuitive. Resilience sounds as if it must be a good thing in principle, but it can prove otherwise in practice. Here is the classic statement about resilience offered by the Canadian ecologist C. S. Holling in 1973. Note that Holling juxtaposes resilience to stability:

> It is useful to distinguish two kinds of behavior. One can be termed stability, which represents the ability of a system to return to an equilibrium state after a temporary disturbance; the more rapidly it returns and the less it fluctuates, the more stable it would be. But there is another property, termed resilience, that is a measure of the persistence of systems and their ability to absorb change and disturbance and still

> maintain the same relationships between populations or state variables. (Holling 2010: 37–8)

A resilient natural system, according to the model first proposed by Holling, will rebound from a disturbance like flood or clear-cutting and fire, yet may not re-establish its former regime of species, instead favouring a new one (or at least one that seems novel to human observers), which may persist for some time in its turn. A resilient natural system therefore has the potential to reach a variety of only *relatively* stable configurations. Among the variety of relatively stable configurations a resilient but perturbed natural system is likely to favour are those that include such unwelcome developments – from a human point of view; the system simply takes and uses whatever it can get – as a higher proportion of invasive and other 'pest' species among its components (like the bark beetles that occasionally decimate yellow pine plantations in the south-eastern US).

The important point to grasp here is that the systemic attributes of the resilient system remain more or less intact, however altered its individual components may be. The system has bounced back, but at what appears (to a human observer) to be an obtuse angle, thanks to the alterations in its appearance. In this instance, *the more things remain the same, the more they can change,* in a reversal of the familiar paradox. And this is the case because the attribute of resilience is distributed throughout the system; that is, as a property resilience is systemic, and therefore it need not be particular – or partial – to a single species or genus. Nor, for that matter, need it favour, within limits determined by geography and climate, one association or community of plants and animals over another.

It appears, then, that stability and resilience are interrelated phenomena in that each is somewhat at odds with the other, since resilience has the potential to create 'bad' forms of stability. Human beings prefer ecosystems that they take to be stable, which is why the concept of stability dominated ecological theory for so long. Resilient ecosystems are more difficult for us to cope with; they require what has come to be called 'adaptive management' and they can play havoc with our expectations about land use. For example: the resilient salt marsh will still be a salt marsh once the hurricane has passed along the eastern seaboard of the US, but it may be dominated by alien *Phragmites* reeds instead of native *Spartina* grasses – and it therefore will be a less welcoming habitat for birds like rails, whose numbers have declined owing to the persistence of dense beds of *Phragmites* in the marshy habitats they favour. So managers of wildfowl refuges along the coast find themselves in something of a stalemate in their attempts to eradicate, or adapt to, the invasive reed. Another example: the resilient savanna will still be a savanna after the cataclysmic bush fire and the prolonged drought. But after the disturbance the savanna may no longer be as dense with woody shrubs and trees, and it therefore will be a better habitat for wildebeest than for elephants. The latter will have to adapt by moving on. The point is that resilience is rather like a casino: it cares less, so to speak, about individual winners and losers, and more about the average take that the house enjoys over time.

So one might say that while the threat of collapse – which in our time is posed most dramatically by global climate change – makes achieving sustainability seem to be an absolute necessity, the reality of resilience suggests that there is some middle ground between the two extremes, a dynamic space where something like an ongoing negotiation over the seemingly stark differences between collapse and sustainability can occur. Juxtaposition of worst-case (dystopian) and best-case (utopian) scenarios of the environmental future has obscured this middle ground in popular discourse, as has the lingering belief that, left to its own devices, the natural world prefers stability. Yet this does not mean that a greater attentiveness to resilience will be the panacea for our environmental woes. The middle ground is contested ground, and while resilience can be a measure of the abiding strengths of natural systems, it can also result in new environmental woes in its own right (such as a preponderance of invasive *Phragmites* reeds, and the further decline of megafauna like elephants that I hinted at above).

Science fiction, speculative fiction and the pre-posterous historical novel

That all four of the terms I just spent some time defining are marked, in varying degrees, by ambiguity underscores their structural importance to the narratives in which they are employed as tropes, owing to a phenomenon readily understood by literary and cultural critics, if perhaps too complacently accepted as routine and unexceptional. The phenomenon has to do with the uncertainty of literary form, especially when it comes to the novel, where it often seems that genre conventions are no sooner put in place than they are violated, vitiated and contravened; that novelistic form is ephemeral, never realised in equipoise but always existing only a hair's breadth from formlessness; and that beginnings, middles and ends can never be as distinct as their idealisation suggests they should be. This unruliness, this kicking over of the traces of convention and this refusal of narrative to move forward along clear-cut lines, makes it possible to read Atwood's MaddAddam trilogy not as a work of science fiction or 'speculative fiction' (the latter is Atwood's preferred term; see the feature article by Potts 2003), but as a tripartite historical novel.

I would argue that such a reading also makes available a better understanding of the roles that collapse, stability, resilience and sustainability play in shaping Atwood's environmental metanarrative, which is mostly implicit but is occasionally expressed in snippets of narration. Admittedly, reading the trilogy as a work of historical fiction will require some sleight of hand if it is going to work. It will have to negotiate, among other things, the awkward fact that Atwood's novels are set in a post-climate change future whose relationship to the present cannot be determined according to the usual measures of chronological succession: if the narrative is not time-stamped and its temporality is uncertain, then so too must be its very historicity.

There is a way to understand this awkward fact about the MaddAddam trilogy in theoretical terms that will make it a useful fact for the reading I propose. We might, like Dipesh Chakrabarty, blame the slippery historicity of Atwood's trilogy on climate change itself. In his landmark essay 'The Climate of History: Four Theses', Chakrabarty observes that 'our historical sense of the present' has become 'deeply destructive of our general sense of history' (2009: 197). Atwood's trilogy documents that destructiveness on every page, as I will show when I discuss the mindset of her protagonist in *Oryx and Crake*. It also will help us to understand the historical character of Atwood's trilogy if we take a long and somewhat jaundiced view of the relevant chapters of literary history. We have to acknowledge that in even the most classic of historical novels – the ones, that is, which are cited regularly in definitions of the genre – historicity does not provide an anchorage in time and place, except in the most general terms. Indeed, historicity is often one of the things such a novel takes not as foundational and therefore for granted, but as problematic – as something to be established, or at least explored, in that novel's text.

This treatment of history as problematic, as less a matter of fact than a mode of inquiry which is necessarily self-reflexive and 'historiographical', has implications that range beyond the confines of the literary (whatever those confines are taken to be; I would not presume to delineate them here). In an essay that takes up the subject of the mutability and fluidity of genre, Hayden White notes that the realist writers of the mid-nineteenth century effectively decided 'to treat "the present" as history', and that 'this move accomplished a metamorphosis of the genre of history writing itself, a change of its focus on the past alone to a focus on the present (and future) of historical societies as well' (2003: 599). White adds: 'Mixture, hybridity, epicenity, promiscuity – these may be the rule now' (2003: 602). In short, historicity itself is not what it used to be, and perhaps it never was; it, too, has always already had a history.

In another essay, White begins by contrasting historical and fictional discourse. The former is interested in the true, he says, the latter in the real: 'The real would consist of everything that can be truthfully said about its actuality plus everything that can be truthfully said about what it could possibly be' (White 2005: 147). Its orientation towards both actuality and possibility (or in Atwood's terms, its 'speculative' character) gives fictional discourse some advantages. Presumably, it allows fiction to embrace, among other things, 'mixture, hybridity, epicenity, promiscuity' as formal strategies (regardless of whether it embraces those topics thematically, or features – let us say – racially mixed, polyglot, promiscuous epicenes as characters). Fiction's radical openness to possibility also gives it another advantage over contemporary historical discourse. Unlike present-day historiography, which desires to have the status of a science, according to White, and therefore cleaves to the past which is its only canonical source of evidence, fiction can continue to ponder the future in much the same imaginative fashion that earlier historiographies influenced by millenarian and apocalyptic Christian thought once allowed themselves to do, or so White argues (2005: 156–7). In yet another essay, this one on Kermode's *The Sense of an Ending*, White glosses Kermode's engagement with myths of beginnings and myths of endings in terms that can be brought to bear directly on the MaddAddam trilogy: 'Such myths, of "unremembered" (and "unknowable") but "imaginable" events, allow us to join an *imagined* beginning with an *imagined* end which pro-retrospectively, that is to say, pre-posterously, endows the time between beginning and end with meaning'. Theological or metaphysical meaning makes for the certitudes of myth and mythology, whereas meaning which seems '*only* imagined or feigned' – meaning which is portrayed as 'only possible' (White 2012: 45, author's italics) – is properly fictional.

Here White offers us a more precise way of characterising Atwood's fictional strategies in the MaddAddam trilogy than calling those strategies 'speculative' (which speaks more to Atwood's unease about being identified as a writer of science fiction than to anything distinctive about the three novels). In the trilogy, we find our own time depicted in *pro-retrospective* and *pre-posterous* ways. Just as Kermode says fiction should, the three novels help us make sense of the 'here and now' (Kermode 1967: 39). And this is what renders Atwood's three novels almost immediately legible: while reading the first of them for the first time, one does have to acclimatise– so to speak – to the world Atwood depicts. But doing that does not mean a wholesale rejection of the norms shaping the world one already inhabits, or the exchange of those norms for new ones. In *Oryx and Crake*, the climate has altered, very much for the worse, and the woods are full of transgenic pigoons, swine endowed with human neocortex tissue thanks to a gene splice. Yet anyone who has experienced a heat wave or seen a pig, and who has some human neocortex tissue of their own, is able to make sense of this world in more or less familiar terms, and by invoking concepts like *hot, humid, mammal* and *pork*, which have been employed for thousands of years. Here and there, the trilogy may be preposterous in the usual sense of the term: in the third volume, for instance, the pigoons develop the ability to communicate telepathically, which is either the unintended consequence of a gene splice too far or a sign that Atwood has lapsed from speculation into science fiction after all – or both. I am assuming that the porcine telepathy does not reflect the fact that the trilogy has jumped tracks entirely and become a work of fantasy fiction, like some other trilogies I could name. Yet this preposterousness of the usual sort does not mean that the trilogy's readers have to do the equivalent of learning how to get by in colloquial Klingon, or brushing up on the doings of the Time Lords who dwell on the planet Gallifrey, to comprehend its meanings.

Apart from these theoretical considerations having to do with genres, conventions and their effects in both the real world and the other worlds posited by fiction (and, inevitably, by interpretations like this one), I want to suggest that a reading of the MaddAddam trilogy as a series of interlinked historical novels, or as a single grand narrative telling the story of the rapid decline and fall – the collapse – of contemporary industrial civilisation, can be achieved without any undue fudging of the details Atwood presents. Consider the following elements of her narrative, each of which is constitutive:

- Climate change, and an accelerating degradation of environmental conditions broadly speaking, resulting in numerous extinctions
- Corporate development of genetically modified organisms, including animals, plants and viruses
- Advanced computer technologies, especially online technologies that help to further the saturation, if not the outright capture, of culture by electronic media
- The accelerating erosion of public space and the emergence of large private compounds owned by corporations; a corresponding decline in the importance of centralised government authority at all levels
- The privatisation of all functions once performed by governments, but especially those functions associated with maintaining infrastructures, ensuring security and upholding standards of public education
- Increased decadence, evident in the popularity of wildly violent and nihilistic video games, the globalisation of the sex trade both online and off, the legalisation of prostitution and recreational drugs, and a precipitous decline in culinary standards so that eating highly processed or entirely artificial foods becomes the norm
- Increased activity on the part of underground environmentalist insurgencies, most importantly the MaddAddam group that gives the trilogy its title
- A global pandemic that decimates the human population, even as some animal and plant species thrive despite the extinction of numerous others.

This list may not be exhaustive. But it highlights most of the things that have been noticed by critics responding to the novels as they were published, and it will serve as a starting point for consideration of the trilogy's historical character in more specific, more concrete terms than I have used heretofore. Please note that many of the items on this list are relevant not only to the future Atwood depicts, but to the present day. The world she imagines is familiar; it is, in almost every respect, our world, and her depiction of it never departs from the norms of realism.

Before turning to the text of the trilogy and discussing it in greater detail, I would like to suggest – really, to insist – that Atwood's approach to her material is satirical. Despite the grimness of many of the elements of her story, which documents the end of the world as we know it though not the end of the world *as such*, Atwood's attitude remains consistently irreverent. While this lack of reverence is basic to the satirist's fictional mandate, it does pose a significant problem, especially for ecocritical interpretations. It makes it difficult to read the MaddAddam trilogy as a cautionary tale about collapse, and inadvisable to try and glean a hopeful, utopian message from the trilogy's treatment of resilience, stability and sustainability. If Atwood's corporate henchmen and boy-wonder scientists are appalling in their reckless disregard for planetary wellbeing, most especially for animal and human rights, they are also whip-smart and quick to counter flabby arguments about the sacred nature of species, or – as they would insist – genomes. Conversely, her environmental activists (many of them corporate renegades themselves) can be almost as hard to stomach as their antagonists, thanks to their constant recitation of pieties about nature's inviolability that came to seem passé some time ago. These activists compromise and even conspire with their antagonists, and some of them offer a green exegesis of the Bible just as strained as anything a late-night televangelist might dream up to justify his belief in, say, faith-healing. The activists also like to perform ecologically minded hymns that are tedious to read, and which even the best of gospel choirs could not render credible as song. Atwood's even-handed approach to her antagonists and protagonists – and it is not easy to sort them out neatly, thanks precisely to Atwood's even-handedness – makes the trilogy a discomforting read. At her best, Atwood takes full advantage of the satirist's mandate to expose contradiction, hypocrisy and lazy moralism, and writes about the good, the bad and the ugly with much the same savage glee.

Oryx and Crake: the importance of being resilient

It seems appropriate to consider the first novel of the trilogy apart from the other two, and to give it priority in my reading, since *Oryx and Crake* is tasked with establishing the character of the fictional world in which all three novels are set (it is also the deftest instalment of the trilogy). As is generally the case with science fiction – and disregard for a moment

Atwood's rejection of that label for her work – as the trilogy opens, the reader has to ascend a learning curve while attempting to absorb the details of what appears to be an alien environment shaped by some new trends in natural history. What makes negotiating the learning curve mildly challenging, at least initially, is that we only learn the things we need to know about this environment piecemeal. To read the novel is to become culturally and environmentally literate, as if the reader also needs to adapt to the transformed world Atwood describes. But for the reader this adaptation is far from being a Darwinian process (we have that advantage, at least, over the pigoons).

Oryx and Crake begins as Snowman, who figures as the main character in this first instalment of the trilogy, wakes just before dawn near an unnamed shoreline:

> On the eastern horizon there's a greyish haze, lit now with a rosy, deadly glow. Strange how that colour still seems tender. The offshore towers stand out in dark silhouette against it, rising improbably out of the pink and pale blue of the lagoon. The shrieks of the birds that nest out there and the distant ocean grinding against the ersatz reefs of rusted car parts and jumbled brick and assorted rubble sound almost like holiday traffic. (Atwood 2004: 3)

The passage presents its reader with something of a perceptual puzzle. The juxtaposition of 'rosy' and 'deadly', followed by the narrator's wistful comment about how 'that colour still seems tender', signals that something is amiss here. Out of place, too, since the towers are situated 'offshore' and just beyond a lagoon filled with what would appear to be tropical waters. The setting might be some coastal city somewhere in the Global South, one ruined by poor or nonexistent urban planning, then abandoned and given over to the shorebirds that now use it as a roosting place. But, as soon becomes clear, the scene, while it is certainly tropical, is only recently so. Most of the action in *Oryx and Crake* transpires on the eastern seaboard of the US, most likely along the stretch of coastline in and around what is now Boston (or so Atwood has indicated in interviews). The first chapter of the novel ends with Snowman eating a mango, one he must have picked himself. Evidently, then, the novel is not set in the present day, but in some future that Atwood is imagining for us as the novel unfolds. In that future, mangoes can be picked well north of their present-day range, and one of the more dire forms of environmental collapse has long since occurred: the climate *has* changed.

It is unlikely that mangoes will manage the move to the Massachusetts coast in time to avoid extinction. However, this implausibility may not matter, since climate change is taken for granted in *Oryx and Crake*. Most of the novel's focal characters are too young to have witnessed the change and have lived with its results all their lives. For them, climate change is essentially a closed chapter of history; this means that it scarcely needs to be *narrated* at all. Here is Atwood's summary account of its progress early in *Oryx and Crake*: 'time went on and the coastal aquifers turned salty and the northern permafrost melted and the vast tundra bubbled with methane, and the drought in the midcontinental plains regions went on and on, and the Asian steppes turned to sand dunes' (2004: 24). This is closer to a list – or to a schoolchild's hastily assembled report, in which all the facts have been cribbed from an old encyclopedia – than to a fully realised passage of narrative. It is offered not because it is interesting and important as such, but merely as a sort of sidebar in a passage about the difficulty of satisfying the craving for fresh meat in the post-climate change world. Attempts to read the trilogy as climate change fiction, and there have been a number of such attempts, overlook the fact that climate change is not disruptive but constitutive of the world that Atwood's characters find familiar, and in which they all seem to be more or less comfortable (at least initially).

Some of Atwood's characters are even a bit bored by the recent history of environmental devastation and collapse that is their ironic birthright. Here is how Snowman, or rather Jimmy since that is his real name, recalls the note his mother left him when she abandoned her family, and a comfortable life in the posh compound of a corporation devoted to transgenic research and development, in order to join an underground environmental group (as Jimmy later learns):

> *Dear Jimmy*, it said. *Blah blah blah, suffered with conscience long enough, blah blah, no longer participate in a lifestyle that is not only meaningless in itself but blah blah.* She knew that when Jimmy was old enough to consider the implications of *blah blah*, he would agree with her and understand. (Atwood 2004: 61; author's italics)

In the original note, the blanks marked by repetitions of the word *blah* were filled with environmental discourse. This discourse fell on deaf ears when Jimmy was young, and by implication, it still does – even with the hindsight Jimmy has gained in the wake of a truly devastating collapse, a global pandemic in comparison to which climate change, to judge from the novel's rather scanty portrayal of it, appears to have been quite manageable.

It is possible to view the circumstances I have just described in diametrically opposed terms, by taking advantage (heuristically) of the dialectical model I sketched in my opening comments about collapse, resilience, stability and sustainability. What has happened before *Oryx and Crake* begins might be characterised less as a collapse than as a failure, a falling short, of sustainability – of precisely the kind we are experiencing at the present moment in history. On this reading, the novel documents something comparatively undramatic: it shows that where climate change is concerned, the present-day habit of balancing prognostication with procrastination and prevarication, the habit of 'debating' climate change, is likely to continue, until all the relevant issues – the need for renewable energy, the shortcomings of industrial manufacturing and agriculture, the rearrangement of urban landscapes, the increased frequency of heavy weather – are obviated by the passage of time and the willingness of human beings to maintain an unsustainable status quo. On such a reading, *Oryx and Crake* is neither apocalyptic nor post-apocalyptic, but 'historical' in the sense of its being self-reflexive and 'historiographical' in the manner I described earlier, when I discussed some ideas formulated by Hayden White. The culture the novel depicts (and the rest of the trilogy follows suit) is not one that has prepared itself mentally for the judgement on its history now being meted out. It is one that has backed resolutely – and *pre-posterously* – into the future with its eyes closed, a culture that continues to 'deny' climate change *even after it has happened*, and which takes only minimal measures (like bullet trains and 'solarcars') to adapt to its ravages.

Jimmy's inability to cope with his situation in the novel's opening pages, like his earlier rejection of his mother's environmentalist message, mirrors his culture's failure to engage adequately with its historical moment. That nobody is able to grasp the enormity of climate change, or to get any real traction on coping with it once its reality is painfully evident, is what makes Atwood's satire so pointed. She describes the systemic failure of local ecologies and reliable weather patterns, and parallels those things with the systemic failure of the culture at all levels. Let's be clear about whose culture she is describing. One of the confessions Atwood might have made in interviews, but to my knowledge has not, is the Flaubertian admission: 'Jimmy is me'.

Perhaps the most striking thing about *Oryx and Crake* is the density of detail with which Atwood fills its pages, but without engaging in a lot of flabby description and exposition (both of which mar the rest of the trilogy). Instead she introduces novel creations like 'Sveltana No-Meat Cocktail Sausages' (Atwood 2004: 4) casually, and tosses off commercial and other neologisms as if they already belonged to the vernacular and should be easily understood. Her brand names are especially amusing, though they are in need of some decoding. The sausages made by Sveltana must be dietetic (surely a first), even if the label does suggest a pun on 'Svetlana', a Russian name which hints that the sausages are marketed to a *babushka* who only dreams of becoming svelte. So the pun suggests advertising images of before-and-after. 'No-Meat' adds to the implication that these sausages are more slimming, and therefore healthier, than other cocktail sausages; though it also indicates that there is literally no meat – or more to the point and more precisely, no longer any meat not originating in the living flesh of a non-GMO – to be had. So the Sveltana cocktail sausages can take their place in the end-of-the-world larder next to the dog food Mel Gibson eats in *Mad Max* and the canned hams scavenged by the protagonists in Cormac McCarthy's *The Road*. Thanks to an accident of history, the Sveltana sausages have become survival rations. As has the 'chocolate flavored energy bar scrounged from a trailer park' which Snowman 'can't bring himself to eat' since 'it might be the last one he'll ever find' (Atwood 2004: 4). Any reader who frequents the food aisles of groceries and convenience stores will be struck by the gooey symbolism of the last energy bar on earth serving, along with the Sveltana sausages, the abandoned skyscrapers and the reef of automobiles that shelters the lagoon, as an ironic monument to the passage of the consumer society we now inhabit.

The norms and parameters of the new world Atwood has imagined take some time to emerge fully. It hinders this emergence only slightly that Snowman is incoherent, perhaps even demented, since his incoherence and dementia coincide with and reflect the alteration in norms and parameters. In the opening pages of *Oryx and Crake*, he speaks aloud several times, once to the grasshoppers he disturbs when, in order to urinate, he climbs down from the tree where he sleeps, and then again to himself, when he says, probably quoting from an old book, 'It is the strict adherence to daily routine that tends toward the maintenance of good morale and the preservation of sanity,' and then again when he begins a sentence he cannot finish: '"In view of the mitigating," he says' (Atwood 2004: 4, 5). Throughout the novel, Snowman is grieving; but he is also borderline aphasic, suffers from auditory hallucinations – and is sometimes drunk to boot. Worst of all, Snowman no longer has anyone to talk to who understands his loss. He is apparently (though not really) the sole human survivor of a global pandemic, so his grip on language and on meaning itself is slipping (which is especially significant since, as we later learn, he is a liberal arts graduate). Of course Snowman is physically miserable, too: dirty, stinking, bitten by bugs and scratched by vegetation, and constantly at risk of sunburn, blindness, dehydration, and lightning strikes thanks to the changed atmosphere, which affords little protection from the sun's rays and is roiled by violent thunderstorms every afternoon.

What makes this situation bearable for the reader is the distance from Snowman's predicament created by Atwood's unfailingly jaunty and often salacious sense of humour, which serves to give the novel perspective on the events that it narrates and the scenes that it describes. It is this perspective on events – and *on the real*, to recall Hayden White's distinction between the historical and the fictional – that makes the novel and its two sequels eligible as works of historical fiction. As events unfold and scenes are developed, the reader is made conscious of the cultural failures of the present moment, and quickly comes to see how a world of just the sort Atwood has imagined might emerge from this moment. In other words, the reader is enabled to see the present moment in parallax view, simultaneously juxtaposed to and synchronised with the future. And this gives rise to a vertiginous sense of movement, of a headlong rush towards a weird future that is all too easy to recognise in the lineaments of the present day, that is already emergent, even as we would like to believe it might be forestalled.

It is for this reason, I think, that Fredric Jameson identifies Atwood as a science-fiction writer (approvingly) and argues that 'at this moment, all fiction approaches science fiction, as the future, the various futures, begin to dissolve into ever more porous actuality' (Jameson 2009: 7). Jameson's point, which he has made many times over the course of his career, is that genre-melding is itself a historical process and is therefore one Atwood cannot escape (her preference for the label 'speculative fiction' notwithstanding). Richard Posner, in a review of *Oryx and Crake*, takes Atwood's measure in more conventional terms, aligning her with Wells, Huxley and Orwell, and suggesting that all four novelists have produced 'extravaganzas of extrapolation' which 'identify a dominant contemporary trend and explore the ominous consequences of its being allowed to continue unchecked' (Posner 2003: 31). Posner's appreciation of *Oryx and Crake* is more restrained than Jameson's appreciation of its sequel, as Posner finds her portrayal of 'today's United States' to be 'a caricature – how much of one is the question' (2003: 32). Yet Posner concludes his review by noting, 'It is increasingly difficult to imagine feasible solutions to the problems created by the scientific-technological juggernaut – the problems dramatised by *Oryx and Crake.* We must not forget that it is in the nature of prophecies of doom that all but the last are falsified' (2003: 36). To put Posner's point into my own terms, and to reiterate an important point I made earlier: just because a certified 'collapse' has never occurred does not mean one will not occur sooner or later. It *can* happen here.

Transgenic, yet all too human

In a situation of the kind Atwood describes – one in which collapse is not so much a dread possibility as a constitutive element of the situation itself, an endemic condition and not just an ominous sign of the times – only a meta-solution will serve. In *Oryx and Crake* Jimmy's boyhood friend Crake provides just such a solution. Throughout the trilogy, Crake remains a cipher, a remote character whose motivations remain unclear despite the lines of dialogue Atwood feeds him, and notwithstanding the background as a disaffected teenager with which she has provided him (in order to hint that while he may be a sociopath, bad parenting and a decadent society are partially to blame). At the same time, Crake gives Atwood a chance to display her own cleverness as a novelist while characterising Crake's intelligence in superficial and shorthand ways (for instance, by identifying him as a numbers guy). As teenagers, Jimmy and Crake spend a lot of time together, smoking pot, watching Internet porn and online broadcasts of capital punishment, and playing computer games. The porn sites – Tart of the Day, Superswallowers and HottTotts – demonstrate Atwood's knack for echoing the facile, and almost always comic, logic of contemporary commercial language. So do online execution sites like shortcircuit.com and brainfrizz.com.

The most formative influence on Crake's development is the computer games he plays with Jimmy, which include *Barbarian Stomp, Blood and Roses, Three-Dimensional Waco, Kwiktime Osama* and *Extinctathon*. Each of these games involves scenarios of collapse, as the names indicate; and each of them, along with the porn and other websites visited by Jimmy and Crake, has been identified, perhaps too readily, by readers as a symptom of the decadence that is one factor leading to the 'post-apocalyptic' conditions described in the novel (see, e.g., Bouson 2004). Atwood herself seems remarkably nonjudgemental. She describes the violent nature of several of the computer games in scandalous detail, and her bemused tone never falters. Here is what she has to say about *Blood and Roses,* where players compete by trading historical atrocities for epoch-making cultural and scientific achievements:

> The exchange rates – one *Mona Lisa* equalled Bergen-Belsen, one Armenian genocide equalled the *Ninth Symphony* plus three Great Pyramids – were suggested, but there was room for haggling. To do this you needed to know the numbers – the total number of corpses for the atrocities, the latest open-market price for the artworks; or, if the artworks had been stolen, the amount paid out by the insurance policy. It was a wicked game. (Atwood 2004: 79)

The note of approval sounded by Atwood's use of the word 'wicked' here should not be missed. *Blood and Roses* may appal the moralist, but its educational value is undeniable. It marries the admiration of monuments beloved by affirmative culture with the levelling strictures of demystifying cultural critique, and thus it might offer something to the Matthew Arnold as well as the Theodor Adorno or the Walter Benjamin in each of us. Here, every document of civilisation can be exchanged, quite directly, for a document of barbarism – and vice versa, too, or Blood-for-Roses if you like. Crake and Jimmy acquire a liberal education by playing the game, and each becomes a more critical thinker as a result. Or so Atwood suggests. (It says something about the redemptive way in which most novels continue to be read and received that while the hymns Atwood wrote for *The Year of the Flood* have been set to music by a composer from California (see the CD *Hymns of the God's Gardeners* (Stoeber 2009)), not one of the much more imaginative computer games she describes in

Oryx and Crake has gone into production. They remain purely literary artefacts.)

While Jimmy is merely disaffected, like all teenagers (he becomes fond of using the dated expression 'bogus!'), Crake's critique is eventually articulated in terms of his contempt not only for his culture but also for his species, both of which he sees as unsustainable. Thanks to his status as a grand master player of *Extinctathon*, and a whiz-kid gene splicer, he acquires the numbers he needs to back up his critique and the skills to render it fully operational. That is, he succeeds in wiping out all of humanity except for a small handful of survivors. (*Oryx and Crake* ends just as Jimmy is about to confront three more of them; others turn up in *Year of the Flood* and *MaddAddam*.) So it is worth noting that at no point in the trilogy does Atwood suggest that Crake's diagnosis of the problems created by his fellow human beings is *wrong*, however ill-advised and mean-spirited his prescription for treating those problems seems to be. This prescription comes in the form of BlyssPlus, a sexual enhancement drug more like the street drug Ecstasy than Viagra™ or Cialis™. Once consumers have become hooked on BlyssPlus, Crake uses the drug as a Trojan horse to introduce a virus, 'a rogue hemorrhagic' (Atwood 2004: 325), at all points of the compass globally and simultaneously. Almost everybody dies.

Thus Crake comes up with a market solution to the problems of collapse and sustainability, but not one of the sort favoured by business and government today, because it is also a meta-solution. All markets crash in the wake of the plague Crake unleashes, as the consumers those markets would like to target liquefy on street corners and in their own homes. The only person Crake intentionally spares is Jimmy, not for sentimental reasons but because Jimmy is the employee Crake entrusts with the caretaking of his greatest creation, the Crakers. They are a new species of transgenic humanoids adapted (if that is the right word) to a diet of crude vegetable matter, obviating the need for agriculture. The Crakers also can eat their own solid waste (a trait borrowed from rabbits), and they can purr (a trait borrowed from cats), enabling them to heal wounds with ultrasound. The males urinate both to relieve their bladders and to mark the boundaries of their territory (a trait borrowed from canines and other predatory mammals). The Crakers also enjoy mating seasons like those of many non-human mammals, so that female Crakers are only 'in heat' a few times a year. The females emit pheromones as their genitals begin to turn bright blue, eliciting a corresponding chromatic change from the males. Since the Crakers are already particoloured (solving the problem of racial difference), their mating rituals give Atwood a chance to describe a scene that resembles a performance of the Chippendale dancers, if one were to attend the show after dropping a few tabs of one of the livelier hallucinogens:

> Courtship begins at the first whiff, the first faint blush of azure, with the males presenting flowers to the females ... At the same time they indulge in musical outbursts, like songbirds. Their penises turn bright blue to match the blue abdomens of the females, and they do a sort of blue-dick dance number, erect members waving to and fro in unison, in time to the foot movements and the singing: a feature suggested to Crake by the sexual semaphoring of crabs. (Atwood 2004: 165)

The mating season is more than the occasion for an amusing spectacle. It solves the problem of sexual jealousy, since each female selects four of the males, all of them physically perfect and very well hung, and mates with them serially until she conceives at the conclusion of what Atwood describes as 'an athletic demonstration, a free-spirited romp' (2004: 165). The mating season also solves the problem of overpopulation, a solution backed up by the Crakers' foreshortened lifespan (which means they will never need geriatric care).

While Crake's redesign and customisation of the human genome seems well thought out and largely successful, there are good reasons to think it fails tests other than, say, the moral and aesthetic ones, which might lead some readers to mourn the fact that sex 'is no longer a mysterious rite' (Atwood 2004: 165) and to think that the Crakers are missing out on something. They are annoyingly inane creatures in many respects. Their dialogue, for instance, does not make for lively reading: they address Jimmy with the worshipful refrain 'Snowman, oh Snowman' throughout the trilogy and they are aggressively literal-minded most of the time. So while I think Atwood should be willing to say 'Jimmy is me' – and even 'Crake is me' – I would not want to hear her say 'The Crakers are me' since the Crakers, especially as they are depicted in the trilogy's first volume, are creatures no thoughtful creator should wish to own.

The chief irony of *Oryx and Crake* is that by engineering the Crakers and placing them alongside all the other transgenic novelties that now fill the landscape, and by wiping out most of the human population at the same time, Crake *does not* reprogramme the course of humanoid evolution as he had planned to do, hacking it like the computer games and other digital technologies he tinkered with as a teenager. He merely reboots it. However altered individual genomes may be in the MaddAddam trilogy, evolutionary processes continue to run just as they always have done. The result of Crake's tinkering is not stability, but resilience – in the Crakers' case, of human nature – and a doubtful forecast for sustainability. The Crakers, tutored by Jimmy, begin to view Crake as a god-like figure in the wake of his death (about which they are not told, since Jimmy has killed him), while Crake's consort – and Jimmy's clandestine lover – Oryx (whose throat was slashed by Crake himself) features in their belief system as a minor deity in her own right. The Crakers also quickly acquire other rudiments of culture and are on their way to becoming fully fledged humanists. This is evident not only in their singing, but also in their fashioning of crude icons and their curiosity about both their own origins and the flotsam and jetsam left behind by vanished human beings (hubcaps, piano keys, bleach bottles, a computer mouse). By the third novel, a young Craker has acquired the ability to read and write, and is able to play a diplomatic role as an envoy in the disputes that emerge between the humans, their Craker counterparts and the now-telepathic pigoons, who have been doing some evolving of their own thanks in part to the human neocortex tissue implanted in them by OrganInc. Crake's design protocols are set aside, then, by a partial reversion to the human phenotype on the part of the Crakers. Of course, it does not help matters that the few surviving and still fertile human females find the advances of the Craker men irresistible; three of them are pregnant with children fathered by Crakers at the trilogy's conclusion. These pregnancies ensure that hybridisation is carried forward not by transgenic but by natural means, although the distinction between the two has been weakened and may no longer apply in the short term.

Resilience therefore seems to play a subversive role in these novels. As I suggested earlier, along with stability, resilience might be said to mediate between collapse and sustainability. But the result is, at best, a slovenly synthesis – and the mediation or rather the evolutionary process seems to be ongoing, with no end in sight. As suggested by the second epigraph to this essay, the end of the trilogy is 'not the end' but an opening onto yet more of the same, perhaps in perpetuity. This suggests, as I hinted earlier, that the logic employed by Atwood late in the trilogy is less that of 'speculative' fiction than that of fantasy, which pre-posterously – and preposterously – makes every narrative over into a never-ending story and forestalls 'the end of history' regardless of whether that end is envisioned as hopeful or not. (It should be clear by now that by proposing that we read the trilogy as a work of historical fiction, I was not seeking to rescue it from being characterised as belonging to some inferior, 'low' genre.)

In *MaddAddam*, the trilogy's 'final' volume, humans have weathered the near-total collapse wrought not by global warming but by Crake's 'rogue hemorrhagic' virus. Yet it seems unlikely that their genome will recover its purity, naturalness and vigour in the future. Assuming, of course, that the genome had any purity and naturalness and vigour left before the catastrophe visited upon it in *Oryx and Crake*, when most humans had already been tweaked in various ways, even if most of those tweaks – implants, fingerprint wipes, and so on – were merely cosmetic. As for the landscape, it will continue to be filled with transgenic plant and animal species run riot: the pigoons will go on competing with other predacious species like wolvogs, bobkittens and humans; the understory will serve as home to feral rakunks and glow-in-the-dark rabbits; lurid flowers, knockoffs of tropical species, will perpetually bloom; and day-glo butterflies will help to pollinate them. So while the few remaining human beings, most of them former God's Gardeners (members of the ecological cult whose history is described at length in *The Year of the Flood*) and conscripts in MaddAddam's ecoterrorist counter-conspiracy, attempt to relaunch the project not of modernity but of sustainability, the post-collapse version of sustainable living is just as compromised as the Deep Ecological version of it they practised in *The Year of the Flood*.

The landscape in *MaddAddam* is so wrecked that the survivors have no choice but to continue scrounging among the remains of the very industrial civilisation they once had hoped to forswear for such essential items as soap, toilet paper and the bed sheets they need to protect their skins from the still-damaging rays of the sun. On the one hand, they have seen the end of industrial civilisation; on the other hand, they continue to be dependent on its products. Equally to the point, the artisanal enterprises they have begun threaten to reinvent industrial processes all over again. The paradox of the MaddAddam trilogy as a work of historical fiction, then, is that it somehow manages to have a false front and a false back at the same time: its end marks a new beginning, but this new beginning seems likely to eventuate in calamities similar to those that – or so we are encouraged to assume – first set its narrative into motion. So in the end, and in the final analysis, the MaddAddam trilogy describes both the course of history and a doom cycle.

References

Atwood, Margaret 2004. *Oryx and Crake*. New York: Anchor Books.

Atwood, Margaret 2010. *The Year of the Flood*. New York: Anchor Books.

Atwood, Margaret 2013. *MaddAddam*. New York: Random House.

Bergthaller, Hannes 2010. 'Housebreaking the Human Animal: Humanism and the Problem of Sustainability in Margaret Atwood's *Oryx and Crake* and *The Year of the Flood*', *English Studies* 91 (7): 728–43.

Bouson, J. Brooks 2004. '"It's Game Over Forever": Atwood's Satiric Vision of a Bioengineered Posthuman Future in *Oryx and Crake*', *Journal of Commonwealth Literature* 39 (3): 139–56.

Caradonna, Jeremy L. 2014. *Sustainability: A History*. New York: Oxford University Press.

Chakrabarty, Dipesh 2009. 'The Climate of History: Four Theses', *Critical Inquiry* 35 (2): 197–222.

Diamond, Jared 2005. *Collapse: How Societies Choose to Fail or Succeed*. New York: Viking.

Grober, Ulrich 2012. *Sustainability: A Cultural History*. Trans. Ray Cunningham. Totnes: Green Books.

Gunderson, Lance H. and Craig R. Allen 2010. 'Why Resilience? Why Now?' In *Foundations of Ecological Resilience*. Ed. Lance H. Gunderson, Craig R. Allen and C. S. Holling. Washington, DC: Island Press, xiii–xxv.

Gunderson, Lance H., Craig R. Allen and C. S. Holling (eds) 2010. *Foundations of Ecological Resilience.* Washington, DC: Island Press.

Holling, C. S. 2010. 'Resilience and Stability of Ecological Systems'. In *Foundations of Ecological Resilience.* Ed. Lance H. Gunderson, Craig R. Allen and C. S. Holling. Washington, DC: Island Press, 19–49. Originally published in 1973 in *Annual Review of Ecology and Systematics* 4: 1–23.

Jameson, Fredric 2009. 'Then You Are Them', Review of *The Year of the Flood. London Review of Books* 10 September: 7–8.

Kermode, Frank 1967. *The Sense of an Ending: Studies in the Theory of Fiction.* New York: Oxford University Press.

McAnany, Patricia A. and Yoffee, Norman (eds) 2010. *Questioning Collapse: Human Resilience, Ecological Vulnerability, and the Aftermath of Empire.* New York: Cambridge University Press.

Posner, Richard A. 2003. 'The End Is Near', Review of *Oryx and Crake. New Republic* 22 September: 31–6.

Potts, Robert 2003. 'Light in the Wilderness'. *Guardian* 26 April. www.theguardian.com/books/2003/apr/26/fiction.margaretatwood. Accessed 15 February 2017.

Stoeber, Orville 2009. *Hymns of the God's Gardeners.* CD. Earthly Ark Music.

Tainter, Joseph A. 1988. *The Collapse of Complex Societies.* New York: Cambridge University Press.

White, Hayden 2003. 'Anomalies of Genre: The Utility of Theory and History for the Study of Literary Genres', *New Literary History* 34 (3): 597–615.

White, Hayden 2005. 'Introduction: Historical Fiction, Fictional History, and Historical Reality', *Rethinking History* 9 (2/3): 147–57.

White, Hayden 2012. 'Historical Fictions: Kermode's Idea of History', *Critical Quarterly* 54 (1): 43–59.

8

'The shadow of the future made all the difference': sustainability in Kim Stanley Robinson's Science in the Capital trilogy

Chris Pak

Anthropogenic climate change and the approach of the peak-oil moment has encouraged many to think about alternative energy regimes that would provide a solution to the threat of economic collapse. While there is consensus amongst climate scientists that climate change is happening, contemporary thought about its specificities and solutions is subject to much debate. Fred Polak argues in *The Image of the Future* (1973) that societies shape themselves partly through the utopian potential of the images of the future that they construct. Science fiction (sf) has portrayed a variety of images of the future, from post-apocalyptic narratives of decline to techno-utopian futures and ecotopian images of sustainable societies. These narratives explore many instances of sustainable and unsustainable practices, but issues of energy, oil, water and the extraction of other resources have been persistent themes. Through portrayals of future worlds and societies that explore the embeddedness of individuals and communities in the realities of their physical and socio-political environments, sf helps us imagine sustainability in a multitude of ways: by presenting specific technological innovations that might support sustainability, by exploring cause-and-effect relationships or the complexity of non-linear dynamic feedback systems, by portraying unsustainable practices and societies that should be avoided, and by depicting characters whose lives are influenced by (un)sustainable practices and who reflect upon and navigate these worlds. Sustainability science, futures studies, and sf all engage in different ways and for different purposes in speculating about the future. Sf cannot offer predictions but it can, as Dominic Boyer and Imre Szeman claim, act as 'a forerunner researching the cultural landscape around us and imagining the future relationship between energy and society that we need to strive toward' (2014).

Sf has portrayed a vast array of ecological images of the future. These narratives offer to futures studies an archive for reflection: a resource of scenarios amenable to a variety of analytical approaches and, sometimes, a commentary on the process of future forecasting itself. Ian Miles notes that 'many futurists are sf aficionados ... and that sf often informs their research' (1993: 1). Karlheinz Steinmüller, a physicist, sf author and scientific director of the foresight company Z-Punkt, argues that sf is not aimed at prediction, but he does call sf 'a kind of fictional technology assessment' (2003: 176) and notes that 'SF constructs future scenarios in a similar way to futurology' (178). This should come as no surprise to anyone familiar with either discipline, for the origins of both can be traced back to the same sources: commentators such as I. F. Clarke (1971) and Eddie Blass (2003), for example, locate their origin in utopian thought. Although 'futurology' was coined in 1943 by Ossip K. Flechtheim (Butler 2014: 513), sf writer Jack Williamson argues that H. G. Wells, in his 1902 lecture *The Discovery of the Future* (1913), invented modern futurology and began from that point to depart from writing sf to propagandising for the realisation of his image of the future (McCaffery 1991). Hugo Gernsback founded the first American sf pulp magazine *Amazing Stories* in 1926 and later coined the term 'science fiction'. His first editorial bore the motto 'Extravagant Fiction Today – Cold Fact Tomorrow!', and he praised the predictive power of writers such as Edgar Allan Poe, Jules Verne, Wells, and Edward Bellamy, claiming that '[p]rophecies made in many of their most amazing stories are being realized – and have been realized' (Gernsback 1926: 3).

Williamson argues in 'Scientifiction, Searchlight of Science' that 'SF was futurology, testing new ideas before scientists got around to them' (1928: 435), but has since revised this bold claim to argue that, in sf, the priorities of fiction take precedence (McCaffery 1991). As Andrew M. Butler explains,

> the readerly encounter with SF involves experiences of sublimity or estrangement through its invocation of imagined (future) environments. By contrast, futurology, Futurism, and futures studies are methods of future *prediction*, with varying needs to persuade an audience into taking or avoiding particular actions; it is much more overtly tendentious than SF, advocating for (rather than merely evoking) potential futures. (Butler 2014: 522)

Futurology, or futures studies, is directed towards imagining futures that could be instantiated. Sf, by contrast, is a heterogeneous artistic mode that draws from a wide variety of traditions. While there are examples of sf that fail or do not attempt to imagine the future (time travel narratives and alternative histories, for example, may be more concerned with the past and parallel presents), sf as a mode is oriented towards imagining futures extrapolated from elements of the contemporary world or with modelling images of the future that reflect back upon the present. In *Green Speculations*, Eric Otto explores the shape of a form of radical ecology he calls transformative environmentalism, which combines influences from a diverse range of oppositional politics that emerged since the 1960s: the science of ecology, environmental philosophy, deep ecology, ecofeminism and ecosocialism. Arguing that 'estrangement, extrapolation, and sense of wonder constitute an ecorhetorical strategy for works of fiction and nonfiction whose interests lie in questioning deep-seated cultural paradigms' (2012: 16–17), Otto identifies capitalism and its logic of limitless growth as the agent of this environmental degradation and the target of transformative environmentalism's critique. Sf is not futurology. Williamson argues that '[p]eople, of course, had always been concerned with understanding and predicting the future; but SF writers, relying on Darwinian insights, have been able to construct fictional visions of the future that are much better based' (McCaffery 1991). Sf is concerned with constructing fictional worlds, for which adherence to facts or truth (relative to our understanding of the physical laws of the universe or the present constitution of elements of the 'real world') is not a useful measure for thinking about the mode; the predictions that futures studies posits, however, can be productively assessed in terms of their veracity and efficacy for risk assessment.

Butler and Williamson accede to what is frequently reiterated in sf scholarship and fandom: that sf is not meant as prediction. Ian Miles, for instance, notes that while many sf writers have 'forecasted ... ideas such as nuclear weaponry and spaceflight – and this list could easily be extended – these treatments remained shoddy until their actuality began to be realised' (1990: 85). Stanislaw Lem, an acclaimed writer of sf and philosophical essays, futurological articles and a member of the committee for Poland 2000 (which attempted to anticipate future trends from its vantage in the late 1970s), has consistently explored the limits of futurology in his fiction and non-fiction, warning in his essay 'Metafuturology' (1986) against the over-specialisation of futurology and for its practice in every discipline. Lem's profound critique of the limits of human knowledge threatens to paralyse attempts at imagining any future. Lem, however, continued to write about the future even when he stopped writing fiction. For him, the practice of future speculation remained valuable even when he believed he had nothing further to contribute to sf. His critique of futurology as a discipline should not be taken as a prohibition against thinking about the future, but as a warning that all such images are provisional, pending the irruption of the unknown.

Sf and futures studies are therefore bound in a relationship that sf writer and editor Frederik Pohl calls 'a pretty amiable symbiosis' (1996: 8). Many writers, such as Arthur C. Clarke, Isaac Asimov, Robert A. Heinlein and John Brunner have been hailed as successful prophets of sf. More recently, figures such as Gregory Benford, David Brin, Cory Doctorow, Karl Schroeder and Kim Stanley Robinson have continued to connect these two ways of thinking about the future. Brin and Schroeder in particular are futurists: Schroeder, for example, was commissioned by the Canadian army to write a narrative of future trends in conflict situations. The result, *Crisis in Zefra* (2005), uses the fictional African city of Zefra to explore future technological and strategic military innovations and to assess their potential risk and competition in a part of the world afflicted by resource scarcity. While Andrew Milner notes that sf may be value-free in the sense that the genre does not imply *a priori* political, ethical or aesthetic values, he does insist that it is value-relevant in that specific texts often do speak to these concerns: 'the future story can be used as a kind of futurology. SF of this kind is intended to be politically or morally effective, that is, to be socially useful' (2012: 180). Futures studies is a utopian discipline and an allied formation to sf. Its influence on writers prompts them to connect fiction to a praxis of speculation and scenario building. If futures studies offers scenarios for structuring prediction based on possibility, fictional narratives build worlds based upon an accumulation of contingencies that are driven by the demands of the form. Sf constructs images of the future from the perspective of actors embedded in their fictional environments. The images sf constructs are not models to guide action, but imaginative spaces for testing ideas and values and (in many cases) a vehicle for encouraging socially engaged reflection on a variety of issues. This requires sf to test scientific and historical facts within the space of their fictional environments, but not necessarily to adhere to them. Sf bears a different relationship to their images of the future from that of futures studies, offering a heterogeneous library of narratives to help think about sustainability.

Sustainability science and sf

Like futures studies, sustainability science is based on the extrapolation of the physical parameters of global or local systems and often draws on predictive modelling and scenario building. Physical systems are the object of their study, but so are the practices, behaviours, values and myths of a society or community, along with the ways these orientations affect sustainability practices. Kim Stanley Robinson has consistently imagined ecological futures that address the relationship between politics, society and science, and has explored ideas related to sustainability, climate change, terraforming, geoengineering and biotechnology. Robinson sees the work of understanding the present as 'a mix of historical work and science fictional speculation' (Davis and Yaszek 2012: 189), while Roger Luckhurst notes that Robinson 'has always regarded science fiction as an inverted form of the historical novel' (2009: 172). For Robinson, historical fiction and sf are related modes of imagining the present through constructed images of the past and future, as is attested by his alternate history about a world where the European population is eradicated by the Black Death, *The Years of Rice and Salt* (2003), and his recent prehistoric novel, *Shaman: A Novel of the Ice Age* (2013). The images of the past and of the future in these works of sf are historical constructs that tell us more about their contemporary moment than they do about either the past or the future.

Robinson has long been concerned with sustainable and unsustainable futures. His Orange County trilogy explores three alternate Californias: *The Wild Shore* (1984) imagines a post-apocalyptic pastoral enclave, *The Gold Coast* (1989) a dystopian, high-capitalist 'autopia' and *Pacific Edge* (1990) portrays a sustainable ecotopian future. Robinson's acclaimed Mars trilogy, comprising *Red Mars* (1996a), *Green Mars* (1996b) and *Blue Mars* (1996c), along with its companion collection of short stories, *The Martians* (2000), locates his thinking about sustainability in the extreme environment of a colonised Mars undergoing terraformation. *Antarctica* (1997), informed by a trip to the eponymous continent that was funded by the National Science Foundation (NSF), is set in and around McMurdo Station and involves characters who would later appear in the Science in the Capital trilogy. Robinson's *2312* (2012) extends his thinking about sustainability by imagining a far future society that has terraformed and colonised the solar system. In his Science in the Capital trilogy, comprising *Forty Signs of Rain* (2005), *Fifty Degrees Below* (2007a) and *Sixty Days and Counting* (2007b), he explores the relationship between science and policy in a near-future scenario where extreme weather events – a consequence of a carbon-based energy regime – realise the predicted effects of climate change. In the rest of this chapter I examine how the Science in the Capital trilogy combines 'proleptic realism' and the 'structural comedy' to identify and analyse the problems associated with addressing the climate crisis. I explore how the trilogy considers the radically transformative potential of sustainable alternatives and ask how it accounts for the failure to adequately address climate change in the trilogy.

Science in the Capital

In contrast to the futures he had previously depicted, Robinson chose to situate the Science in the Capital trilogy closer to the time of its writing in a future that Luckhurst identifies with the George W. Bush administration (2009: 171). By aligning without identifying this fictional president with Bush, and the narrative's Senator Phil Chase to Al Gore, Robinson imagines how America might re-orient itself in relation to an alternative, sustainable vision of the future. Adeline Johns-Putra argues that '[t]he dramatic and emotional contours of climate change have to do with the future, not the past or present' (2010: 749), and that Robinson's narrative of abrupt climate change allows him to bring these concerns to bear in a near-future setting located so close to the present that it could stand in for the now. Luckhurst labels the style of those moments where the trilogy remains wedded to a mimetic representation of the present day 'proleptic realism', a contrast to Robinson's previous experiments with ecologically oriented sf that imagines sustainable and unsustainable futures on worlds recognisably different from our own. In this trilogy, speculation about the future is firmly grounded in a fictional world that is clearly and plausibly connected to our contemporary real world, thus bringing it closer – though not completely aligning it with – the project of futurological speculation. This allows Robinson to call into question the assumptions, institutions and practices that retard a movement towards sustainability. Climate change often evokes catastrophic images of the future that might be avoided if alternatives to current polluting practices are instituted. The extreme weather events that are imagined in the trilogy threaten the integrity of the environment and the cohesion of society; it is an image of the future that insistently presses upon the now and calls for immediate action to mitigate its effects. Although it is an important strategy that informs the trilogy, Robinson does not present us with a futurological extrapolation, but with a fictional assessment of the actions that chart a movement towards an imagined future, one that cannot be clearly traced from the initial conditions of the real-world contemporary to the text's publication.

In addition to its proleptic realism, Luckhurst and several other commentators have considered the trilogy's status as comedy, a mode that Robinson also mobilised in his Orange County and Mars trilogies. Robert Markley (2012), for example, points to features such as the concluding marriage between President Phil Chase and the director of the NSF, Diane Chang, as a symbol for the union between politics and science. Markley notes that '[c]omedy invokes generically both a movement towards the restoration of a disturbed social order and the generational continuity typically symbolized by marriage' (2012: 12). As Douglas De Witt Kilgore explains, the trilogy avoids the trope of catastrophe and survivalist recovery, offering instead a 'structural comedy' 'in which the world is reimagined, but preserved in its current social complexity' (2012: 101). Depicting these characters' negotiation of social and political structures and relations allows Robinson to build a fictional world that models one vision of techno-social change, along with the plurality of modes of awareness and agency from which this change emerges. Comedy frames the imagination of possibilities for a restoration of social relations and a movement towards sustainability that allows the narrative to explore the implications of inspiring and enacting change. While narratives involving catastrophe or survivalist recovery focus attention on the conditions of a post-catastrophe environment, Robinson's use of comedy to explore the imbrication of action by individuals directs attention to the sustainable future that the trilogy's characters attempt to instantiate in lieu of a seemingly inevitable apocalypse.

Central to Robinson's structural comedy is the representation of the bureaucratic process, which allows Robinson to explore the values that substrate two broad positions on climate change. The US administration's official position on climate change at the beginning of the trilogy is based on ameliorating both the implications of increased carbon dioxide and the scientific methods used to calculate the impact of high emissions, tracked at 600ppm in the narrative as compared to the real-world figure of 400ppm in December 2014. In an impromptu meeting with the president and his scientific advisor, Dr Zacharius Strengloft, Charlie Quibler finds himself defending the efficacy of measurements by ecological footprint and of acceding to the precautionary principle against Strengloft's accusation that 'those concepts are not good science' (Robinson 2005: 160). Strengloft and the president inconsistently emphasise both debate and appeals to 'good' science in order to maintain the current system of carbon use that underpins the American economy. In defence of his views, Strengloft suggests that '[y]ou need a diversity of opinions to get good advice' (Robinson 2005: 156) – a statement that the administration's actual practice belies. Strengloft's appointment as the president's scientific advisor reflects the administration's desire to replace the previous advisor precisely to eliminate debate; his predecessor's view is that 'global warming might be real and not only that, amenable to human mitigations' (Robinson 2005: 155). In response to Charlie's assessment of the widespread agreement regarding climate change, Strengloft counters with the circumlocution '[w]e've agreed that there is general agreement that the observed warming is real' (Robinson 2005: 159), and he dismisses these indexes by comparing them to less conventional measures of a country's success: '[n]ext you'll be wanting us to use Bhutan's Gross Domestic Happiness' (Robinson 2005: 161). While there are good bases for questioning the measures used to assess economic growth and its impact on society, Strengloft and the administration's assessment derives from a sense of American exceptionalism that makes them subject to different criteria. Strengloft argues that 'we can't use little countries' indexes, they don't do the job. We're the hyperpower,' and he frames 'the anticarbon-dioxide crowd' as 'a special interest lobby in itself', thus marginalising their dominant voice in a debate whose parameters are established in narrow conformity to one image of economics (Robinson 2005: 161).

At the heart of this resistance to adopting the precautionary principle and reducing emissions is the way in which carbon is intertwined with the American economy and its infrastructure: 'everything would have to change, the power generation system, cars, a shift from hydrocarbons to helium or something, they didn't know, and they didn't own patents or already existing infrastructure for that kind of new thing' (Robinson 2005:

156). The energy and transport sectors' reliance on carbon ensures that the larger economic and infrastructural system is so inflexible that any modifications that might impinge upon the interests of business and industry are precluded. Robinson's reworking of comedy, with its attention to the everyday realities of individuals in their techno-social contexts, encourages this focus on infrastructural systems and the social implications of their transformation towards sustainability. Uncertainty about the future is the most rhetorically effective argument marshalled against attempts to modify the current system. The president maintains, 'we don't know for sure if any of that [global environmental degradation] is the result of human activity. Isn't that a fact?' (Robinson 2005: 159). The administration emphasises uncertainty because it provides the economic justification for supporting ineffective measures to address climate change. The system of reason that underlies economic forecasting requires complete certainty in order for business to ground their projections of growth. This simplification of reality mistakes the work of forecasting, which operates not on certainty, but possibility; climate change science offers scenarios or images of the future that are intended to inform current action through an assessment of possibility and risk. The complexity of Robinson's use of structural comedy counters these economic simplifications to explore the social ramifications involved in creating a new image of the future to strive for. The administration's answer to Charlie's competing image of the future is an attempt to restrict the imagination of the future. The president argues that '[y]ou've got to stick to the common sense idea that sustainable economic growth is the key to environmental progress' (Robinson 2005: 165); in other words, it is '[e]asier to destroy the world than to change capitalism even one little bit' (Robinson 2005: 156).

This notion of sustainable economic growth, like many of the devices used to frame the climate debate, is used to redirect the administration's approach to the advice of the majority of organisations and scientists who highlight the environmental and social impacts of climate change. This concept shifts the value of 'sustainable', from one denoting a future where resources are able to circulate within a feedback system as close to indefinitely as possible, to one where expansion becomes the object that must be sustained. It introduces a system of calculation in which the negative costs associated with the use of carbon are consistently measured against the projected profits of the expansion of a carbon-based economic system. Because these calculations are grounded in an assessment of the present costs and profits associated with growth, and because any uncertainty closes debate regarding the value of those costs and profits when projected into the future, sustainable economic growth can effectively be used to delay a constriction of the carbon economy.

Phil Chase, the newly elected president of *Sixty Days and Counting* and a character who echoes Al Gore's position on climate change, offers an assessment of why capitalism is unable to address the climate crisis effectively:

> It isn't the easiest money yet. Capital always picks the low-hanging fruit first, as being the best rate of return at that moment. Maximum profit is usually found in the path of least resistance. And right now there are still lots of hungry undeveloped places. And we haven't yet run out of fossil carbon to burn. Heck, you know the reasons – it would be a bit more expensive to do the start-up work on this country called sustainability, so the profit margin is low at first, and since only the next quarter matters to the system, it doesn't get done. (Robinson 2007b: 378)

Like the crude oil that is pumped along pipelines around the world, profit is imagined as liquid, flowing along the most convenient channels and accumulating as capital for further corporate expansion. In contrast to Robinson's portrayals of sustainable systems on other planets in his Mars trilogy and *2312*, Chase inherits a system whose infrastructure sets conditions on the costs of change, necessitating a recalibration of the meaning of growth and sustainability. History, by way of the ecological and socio-political systems that regulate economic and industrial growth, determines the possible actions that characters are able to initiate. In response to an image of the future to which the past and present are projected, Chase offers a pastoral, utopian image of a 'country called sustainability', a place that has no extension in space and which is arguably eternally deferred (Robinson 2007b: 378). This image draws on the utopianism of a new Eden on Earth, one that promises instantiation through a transformative will and combined effort. In contrast to the US administration's refusal to recognise the climate crisis, this image embeds value into the future and so attempts to orient society towards its creation by offering an image of the potential environmental restoration that could be achieved. Robinson's focus on structural complexity is thus central to the text's project of outlining the considerations necessary to address the correspondingly complex nature of climate change. Yet this image operates as a simplification, as an ideal to continually work towards rather than a warning about the future, and it relies on the possibility of recalibrating present conceptions of economics and growth. It is an image of the future that inspires individuals to engage actively in its creation, and not a prediction based on the initial conditions of the present day. Carbon-based capitalism is motivated by short-term goals and profits; the failure to recognise adequately the current limits to the integrity of this system can be traced to the pre-established imbrication of social, economic and geopolitical networks that exclude many of those countries identified as economically undeveloped from engaging with the decision-making process that maintains the flow of capital.

The failure to widen the sustainability debate to include previously marginalised voices is dramatically represented by the Khembalis, Tibetan Buddhists in exile on an island threatened by rising sea levels. They establish an embassy and become friends and allies of Anna and Charlie Quibler, Frank Vanderwal and Phil Chase, who assist them in their efforts to raise awareness of their plight and the ways in which Buddhism can complement scientific inquiry so as to reach that 'country called sustainability'. Markley suggests that the Khembalis function as 'a chorus for the efforts of Frank, Charlie, Anna, Phil, and Diane as they struggle to resacralize humankind's relationship to a natural world that is very different from the one described by Emerson and Thoreau' and that '[i]t is only through a collective rethinking of history, science, and Nature that a new civilization can begin to emerge' (2012: 10). Kilgore points to another role that the Khembalis play in re-forming the perception of the place of science for society, whereby 'an ethical dimension is claimed for science that goes beyond its role of providing a Verne-like catalog of nature's wonders' (2012: 98). Rather than a system of rationality that supports economic reason, the Khembali ambassador Rudra Cakrin reframes science in Buddhist terms as a way to develop compassion for a species; Robinson, Kilgore argues, 'is calling for its [science's] reformulation from within a regime that requires reason only to be profitable to a new dispensation in which it is part of a more generous common good' (2012: 98–9). In other words, sustainable economic growth is based on a narrow range of beneficiaries – those with interests in the ongoing system of carbon capitalism. By representing their interests and by recontextualising the role of science for society, the Khembalis widen the pool of interests that any attempt to address climate change must take into account.

Reflecting the social complexity of the structural comedy, the Science in the Capital trilogy follows the actions of policy-makers, political advisors and scientists who are engaged in the work of reimagining and recalibrating the relationship between politics and science. It attempts to give voice to these alternatives through the subjectivity of its characters as they negotiate a new landscape brought about by premonitory disasters that index an image of a climate-stricken future. The extreme weather events – the flood at the end of *Forty Signs of Rain*, the cold snap and storms of *Fifty Degrees Below*, and the heatwaves of *Sixty Days and Counting* – bring the image of the future insistently to the fore as the protagonists attempt to formulate policies to avert the worst-case climate scenarios. In *Forty Signs of Rain*, regular blackouts bring the 'shadow of the future' (Robinson 2005: 126) to the forefront; Frank predicts that '[t]his is what it's going to be like all the time ... We might as well get used to it' (Robinson 2005: 328). The possible scenarios that emerge from the present make adaptation inevitable. Anna, reflecting on the future her son Joe would inherit, wonders, '[w]hat was worry, after all, but a kind of fear? It was fear for the future.

And in fact the future was bound to bring its share of bad things, there was no avoiding that' (Robinson 2007a: 245). In contrast to Frank's gloom, she philosophises that worry 'was an anticipation of grief, a nightmare of the future. A species of fear; and she was determined not to be afraid' (Robinson 2007a: 246). The shadow of the future compels change in response to the realities of the physical world. It does not imply a single orientation but evokes both fearful and pragmatic responses that are connected to a generational perspective. Anna refuses to succumb to an apocalyptic image of climate instability; rather she connects her efforts to the optimism and utopian energy embodied by the later president, Phil Chase's, approach to climate adaptation and mitigation.

Charlie attempts to shift the notion of sustainable growth as tuned to the maintenance of a carbon economy by presenting climate rectification and bioinfrastructure mitigation as new industries that offer modes of economic expansion better suited to the meaning of 'sustainability' as 'enduring' and 'capable of long-term maintenance'. In his meeting with Strengloft and the unnamed president at the beginning of the trilogy, Charlie portrays these areas as 'a growth industry with uncharted potential. It's the future no matter which way you look at it' (Robinson 2005: 164). Charlie's policies underpin the public's view of Phil Chase as a prophet of climate change, despite Chase's pragmatic dismantling of Charlie's proposed legislation at the beginning of *Fifty Degrees Below*. When he is elected president and begins in *Sixty Days and Counting* to enact a wide-ranging, environmentally conscious platform, he commits the American people to a programme of sustainability based on three pillars: 'technology, environment, and social justice' (Robinson 2007b: 92). Technological solutions based on clean energy are central to this movement towards a sustainable future, but social justice – especially for women and children around the world – is essential, too, to address exponential population growth and its effects on maintaining a sustainable culture: 'So this is one of those situations in which what we do for good in one area, helps us again in another. It is a positive feedback loop with the most profound implications' (Robinson 2007b: 92). Conceptualising society in terms of these three pillars characterises it as a non-linear dynamic feedback system, a structure that draws on complexity theory and which Robinson aligns with the structural comedy. The trilogy is able to explore the relationships between ecological systems, climate change and climate mitigation by hybridising sf with the structural comedy to portray a fictional future in all its socio-political complexity.

In a series of blog posts that Chase addresses to the American people, he frames this movement towards sustainability as the creation of a permaculture, a dynamic culture that is able to adapt to change but which maintains the goal of long-term sustainability for future generations. Chase's utopian vision of a sustainable country displaced into the future is an attempt to re-orient American values by introducing an element of universality in space and in time; the work of creating a sustainable permaculture is dependent upon assistance to developing countries and an expansion of these values to the globe:

> Eventually I think what will happen is that we will build a culture in which no one is without a job, or shelter, or health care, or education, or the rights to their own life. Taking care of the Earth and its miraculous biological splendor will then become the long-term work of our species. We'll share the world with all the other creatures. It will be an ongoing project that will never end. People worry about living life without purpose or meaning, and rightfully so, but really there is no need for concern: inventing a sustainable culture is the meaning, right there always before us. (Robinson 2007b: 516)

In contrast to carbon capitalism, which depends on unemployment in order to limit wage increases, this sustainable image of the future imagines an ongoing project of social justice, environmental stewardship and a responsibility to future generations. Because it is an ongoing process, the goal of reaching a sustainable future country is eternally deferred. It instead provides a constant motivating image to direct intervention in the present. Chase's platform is informed by the work of the NSF as they re-orient their practices around the project of developing a permaculture. Central to this vision of a feedback system between techno-science, politics and a concern for the biosphere is the project to establish science itself as a political actor embodied by practising scientists. Led by Frank, several members of the NSF establish the 'Social Science Experiment in Elective Politics', or SSEEP, which aims to institute a scientifically informed approach to interventions in society and politics. Fundamental to the Science in the Capital trilogy is this bridging between science and politics and the orientation of this alliance towards the development of a permaculture. Despite the criticism levelled at the NSF for attempting to close the traditional divide between science and politics a divide that Frank traces to the alliance between science and the military during World War II and the formal separation of the military from the political sphere – he argues that '[s]cience isn't like the military. It's the solution, not the problem. And so it has to insist on itself ... we have the only methods there are to deal with these global environmental problems' (Robinson 2005: 325).

The NSF's exploration of sustainable alternatives to the current carbon economy involves a broad-based approach and an interaction between several economic and scientific disciplines, but 'energy was at the heart of their problem' (Robinson 2007b: 237). Edgardo, a scientist centrally involved in drafting the SSEEP, identifies the system of foreign policy that ties carbon extraction, war and the arming of foreign nations and groups into a feedback system that underpins the contemporary global economy: 'we blew the fossil-fuel surplus on wars, and lost the chance to use a onetime surplus to construct a Utopian scientific society' (Robinson 2007a: 530). Rather than using the capital extracted from this system to bootstrap America towards a sustainable society, it was used to sustain an expansion of global markets to further increase the accumulation of capital. The NSF SSEEP committee reflect on an internal document commissioned by Andrew W. Marshall of the Pentagon, 'Imagining the Unthinkable', which outlines possible future scenarios in the event of abrupt climate change caused by the stalling the Gulf Stream (Schwartz and Randall 2003). This real-world Pentagon report exemplifies the use of future scenarios for risk assessment. The impact of this image of the future, however, does not extend to addressing the potential climate crisis. According to Edgardo's assessment,

> The World Bank's Extractive Industries Review Commission had recommended they cut off all future investment in fossil fuels, and move that same money into clean renewables. But in the end the World Bank board voted to keep their investment pattern the same, which was ninety-four percent to fossil fuels and six percent to renewables. (Robinson 2007a: 81)

Their task is clear: in the face of the evidence of abrupt climate change, they must overcome the infighting between agencies with interests in different forms of energy production in order to harness new technologies – 'some combination of sunlight, wind, wave, tide, currents, nuclear, and geothermal power' (Robinson 2007b: 243) – to shift the economy from its carbon base to a flexible portfolio of types and degrees of clean energy.

Much of the NSF's activity to address abrupt climate change involves building an infrastructure that would link disparate research groups engaged in various aspects of the movement towards a permaculture. This top-level organisation of scientific approaches to the climate crisis is an essential element of the NSF's re-orientation towards political intervention. Anna's approach to sustainability is focused on evaluating specific scientific projects aimed at addressing various aspects of the climate crisis, in marked distinction to Chase's broad-based appeal to the American public, yet both are organised around an empowering utopian image of progress towards a sustainable future. Anna delves into the large number of projects that had previously been instituted, but which were unable for several reasons to sustain their inquiry: she 'was finding the fossil remnants of various foreign-aid programs that had been focused on science infrastructural proliferation, as she called it. Some of these were inactive because they were funding starved; others had been discontinued' (Robinson 2007a: 529). Much of the work of developing a permaculture need not be invented *ex nihilo*, but rather a judicious delving into the library of scientific research and infrastructure management, along with the stitching together of a series of approaches in response to the image of the future that Chase inaugurates and that the NSF constructs, offers an avenue for the willed effort that the creation of a permaculture requires.

This delving into the library of past projects is part of an experimental approach to technological methods for addressing climate change. The most extensive series of interventions imagined include several geoengineering projects that bring the principles of terraforming that Robinson explores in the Mars trilogy to Earth. The implications of geoengineering as a form of climate rectification and mitigation – the fact that these approaches offer direct intervention with the climate but are also experimental and potentially disastrous – is made immediate by locating them on Earth. Such large-scale efforts at direct climate mitigation are displaced into the future in the form of proposed research projects, such as Frank's suggestion for the modification of patterns of precipitation by 'flooding the world's desertified lake basins', a task with '[u]ltimate effects [that are] impossible to predict' (Robinson 2007b: 242). One exception is the restarting of the stalled Gulf Stream, unusual for being one of the few problems amenable to direct intervention in this narrative, and in that sense it is 'an anomaly' (Robinson 2007a: 623). Similar to the co-opting of capitalism to bootstrap towards a new energy economy based on renewables, Frank draws on the skills of the American Army Corps of Engineers as one of few organisations with the resources and the capability to enact these large-scale engineering projects. Frank reflects that:

> The world was their sandbox. Castles and moats, dams and bulwarks ... they had drained and then rehydrated the Everglades, they kept New Orleans dry, they had rerouted all the major rivers, irrigated the West, moved mountains. You could see all that right there on the general's happy face. Stewardship, sustainability – fine! Rack but not ruin! Working for the long haul just meant no end, ever, to their sandbox games.
>
> 'No deep ecologists in the U.S. Army Corps of Engineers, I guess.' (Robinson 2007a: 215)

The 2005 Hurricane Katrina and the failure to address its material, social and economic impact in the immediate aftermath of the disaster highlights limits to the army's capacity to mitigate the effects of extreme weather events and adds an additional unintended resonance to the army's effectiveness in future scenarios. Katherine Buse has explored how *Sixty Days and Counting* responds to Hurricane Katrina and the threat that it poses for American conceptions of the Nation, pointing out that 'Americans saw the conditions caused by the storm and recognized the conditions of the Global South, the conditions of the anti-progressive swamp, on

American soil' (2011: 46). Portrayals of landscapes undergoing change highlight the traditional system of value that such environments are made to represent. Transformations to these environments are accompanied by a shift of socio-cultural values that illustrates the networked relationships between culture and the environment. Stewardship and sustainability provide a new rationale for the extension of the view of Earth as a sandbox for manipulation. In *Sixty Days and Counting*, General Wracke explains that '[t]he Corps has always done things on a big scale. Huge scale. Sometimes with huge blunders', and optimistically claims that '[l]ots of things are reversible, in the long run. Hopefully this time around we'll be working with better science. But, you know, it's an iterative process' (Robinson 2007b: 214).

In discussion with the Office of Management and Budget, Frank discovers that, according to their costing, 'they could swap out the electricity-generating infrastructure for about three hundred billion dollars – an astonishing bargain' (Robinson 2007b: 339). More affordable, however, is an image of a geoengineered future based on sustainable agriculture. This project is framed as a new expansion into the American West and one possible answer to rising unemployment: '[i]n other words they needed more cowboys, incredible though that seemed ... [t]he emptying high plains – you could repopulate a region where too few people meant the end of town after town. Landscape restoration – habitat – buffalo biome – wolves and bears. Grizzly bears. Cost, about fifty billion dollars' (Robinson 2007b: 339). Landscape restoration and the rewilding of the American wilderness, along with the possibility of reviving images from a Romantic past, are linked to sustainability. Characteristically, despite the resistance and conflict between government agencies, Frank's involvement in the imagining of new images of the future leads him to discover many possible alternatives to the challenges climate change brings. Phil Chase is central to the reframing of these challenges, not according to catastrophic images of the future, but in terms of a striving towards a utopian goal that is, in Wracke's words, 'an iterative process' (Robinson 2007b: 214). This utopia of process depends on coupling science and politics and thus giving those who practise science the ability to contribute effectively to establishing an informed approach to policy. The goal of reaching that 'country called sustainability' involves a process that Diane summarises as 'first finding bridge technologies, moving away from what they had now while still using it – then the next real thing, the next iteration on the way to a completely sustainable technology' (Robinson 2007b: 237).[1] The past, then, is yoked to an image of the future that gives it new shape, allowing transformations that answer to the needs of a present undergoing severe change. Yet the 'next iteration' towards sustainability cannot be imagined. Instead, the inspiring image of a country called sustainability substitutes a clear path towards that ideal for an image of the future that operates as a guiding principle and a conceptual contrast to contemporary approaches to climate change and sustainability.

Robinson's Science in the Capital trilogy continues the inquiry into environmentalism, ecopolitics, and sustainability that his groundbreaking Mars trilogy engages. Considering the impact of the future on science, society and politics on Earth allows Robinson to reconnect many of the issues explored in that trilogy directly to the infrastructure – the life-support system – of Earth. The Science in the Capital trilogy imagines a future that gestures towards clean energies as a replacement for carbon capital as a foundation for a new economic system. The depiction in this sf narrative of an alternative approach to climate change helps clarify the demands that contemporary images of the future place upon our societies and on us as individuals. It also helps us think around the complexity of sustainability through a narrative form that translates speculation about the future into an imagined experience of living in the immediacy of abrupt climate change and of working towards the goal of sustainability. In many cases, the strategies and policies explored in these texts are stop-gaps, bridges towards sustainability that aim to preserve a life-support system that is basic to nurturing survival but which requires the management and transformation of an inherited economic system based on carbon.

The trilogy depicts many reflections on oil and on ways to sequester and manage atmospheric carbon. These technological considerations are given a utopian cast through the 'Contract with America' that Chase adopts, itself a reformulation of the SSEEP. The blog posts that Chase publishes signify an alliance between science and a socially-minded politics that presents itself as truly accountable to the public. Diane and Frank's growing politicisation over the course of the trilogy allows them to create an infrastructure to deal with climate change through a deliberate overhaul of funding practices and the sifting through the library of old scientific programmes. The 'Contract with America' is a vision of a permaculture that expresses the administration and the NSF's political values. It helps to direct their actions in their efforts to re-orient society toward a sustainable future. The *Science in the Capital*'s approach to sustainability is rooted in the celebratory and fearful responses to societal change that the image of a catastrophically-damaged future biosphere portends. Through an exploration of the social and technological complexity involved in addressing the climate crisis, Robinson does not offer a prediction so much as an image around which society might circle in an ongoing, iterative process of environmentally oriented socio-political transformation. This socially conscious element is symbolised by an image of a future country called sustainability that exceeds strictly national boundaries. It instead mobilises the structural comedy in order to gesture to a future in which society is restored by virtue of a shared movement towards sustainability.

Notes

1 The discourse of transitional, bridge technologies is an important feature of the debate surrounding fracking and represents an application of sustainable images of the future to preserve an untenable carbon-based infrastructure. While the discourse of a transition to a sustainable future is now an accepted element of American governmental policy – as recognised by the COP 21 Paris Agreement – the political, social and scientific approach to developing truly sustainable solutions to climate change outlined in the Science in the Capital trilogy provides an implicit critique of the COP 21 process, which excluded oppositional voices and thus narrowed the range of beneficiaries factored into solutions for climate change. Article 14, which states that assistance should be given to countries in the Global South disproportionately affected by climate change, was shifted to the preamble of the agreement, thus making it non-binding and thus ameliorating the effectiveness of the agreement for tackling climate change at a global level.

References

Blass, Eddie 2003. 'Researching the Future: Method or Madness?', *Futures* 35: 1041–54.

Boyer, Dominic and Imre Szeman 2014. 'The Rise of Energy Humanities: Breaking the Impasse'. *University Affairs*. 12 February. www.universityaffairs.ca/opinion/in-my-opinion/the-rise-of-energy-humanities/. Accessed 20 February 2017.

Buse, Katherine 2011. '"It is Our Fate to Stay Out Here:" America from Swamp to Savannah and Back Again'. Undergraduate dissertation. Durham, NC: Duke University.

Butler, Andrew 2014. 'Futurology'. In *The Oxford Handbook of Science Fiction*. Ed. Rob Latham. Oxford: Oxford University Press, 513–23.

Clarke, I. F. 1971. 'Prophets and Predictors: 1: The Utility of Utopia', *Futures* 3 (4): 196–401.

Davis, Doug and Lisa Yaszek 2012. 'Science's Consciousness: An Interview with Kim Stanley Robinson', *Configurations* 20 (1/2) (Winter–Spring): 187–94.

Gernsback, Hugo 1926. 'A New Sort of Magazine', *Amazing Stories* 1 (1): 3.

Johns-Putra, Adeline 2010. 'Ecocriticism, Genre, and Climate Change: Reading the Utopian Vision of Kim Stanley Robinson's Science in the Capital Trilogy', *English Studies* 91 (7): 744–60.

Kilgore, Douglas De Witt 2012. 'Making Huckleberries: Reforming Science and Whiteness in Science in the Capital', *Configurations* 20 (1/2): 89–108.

Lem, Stanislaw 1986. 'Metafuturology', *Science Fiction Studies* 13: 261–71.

Luckhurst, Roger 2009. 'The Politics of the Network: The Science in the Capital Trilogy'. In *Kim Stanley Robinson Maps the Unimaginable: Critical Essays*. Ed. William J. Burling. North Carolina: McFarland, 170–80.

McCaffery, Larry 1991. 'An Interview with Jack Williamson', *Science Fiction Studies* 54.18 (2). www.depauw.edu/sfs/interviews/williamson54interview.htm. Accessed 20 February 2017.

Markley, Robert 2012. '"How to Go Forward": Catastrophe and Comedy in Kim Stanley Robinson's Science in the Capital Trilogy', *Configurations* 20 (1/2): 7–21.

Miles, Ian 1990. 'Fiction and Forecasting', *Futures* (January/Februrary): 83–91.

Miles, Ian 1993. 'Stranger than Fiction: How Important is Science Fiction for Futures Studies?', *Futures* (April): 315–21.

Milner, Andrew 2012. *Locating Science Fiction*. Liverpool: Liverpool University Press.

Otto, Eric 2012. *Green Speculations: Science Fiction and Transformative Environmentalism*. Columbus: Ohio State University Press.

Pohl, Frederik 1996. 'Thinking about the Future', *The Futurist* (September–October): 8–12.

Polak, Fred 1973. *The Image of the Future*. New York: Elsevier.

Robinson, Kim Stanley 1984. *The Wild Shore*. New York: Tom Doherty.

Robinson, Kim Stanley 1989. *The Gold Coast*. London: Futura.

Robinson, Kim Stanley 1990. *Pacific Edge*. London: Unwin Hyman.

Robinson, Kim Stanley 1996a. *Red Mars*. London: Voyager.

Robinson, Kim Stanley 1996b. *Green Mars*. London: Voyager.

Robinson, Kim Stanley 1996c. *Blue Mars*. London: Voyager.

Robinson, Kim Stanley 1997. *Antarctica*. London: Voyager.

Robinson, Kim Stanley 2000. *The Martians*. London: Voyager.

Robinson, Kim Stanley 2003. *The Years of Rice and Salt*. London: HarperCollins.

Robinson, Kim Stanley 2005. *Forty Signs of Rain*. New York: Bantam.

Robinson, Kim Stanley 2007a. *Fifty Degrees Below*. New York: Bantam.

Robinson, Kim Stanley 2007b. *Sixty Days and Counting*. New York: Bantam.

Robinson, Kim Stanley 2012. *2312: A Novel*. London: Orbit.

Robinson, Kim Stanley 2013. *Shaman: A Novel of the Ice Age*. London: Orbit.

Schroeder, Karl 2005. *Crisis in Zefra*. Department of National Defence: Ontario. www.kschroeder.com/foresight-consulting/crisis-in-zefra/Crisis-in-Zefra-e.pdf. Accessed 20 February 2017.

Schwartz, Peter and Doug Randall 2003. 'An Abrupt Climate Change Scenario and its Implications for United States National Security'. Climate Institute. https://fas.org/irp/agency/dod/schwartz.pdf. Accessed 16 March 2017.

Steinmüller, Karlheinz 2003. 'The Uses and Abuses of Science Fiction', *Interdisciplinary Science Reviews* 28 (3): 175–8.

Wells, H. G. 1913. *The Discovery of the Future*. New York: B. W. Huebsch. www.gutenberg.org/cache/epub/44867/pg44867.txt. Accessed 20 February 2017.

Williamson, Jack 1928. 'Scientifiction, Searchlight of Science', *Amazing Stories Quarterly* 1 (4): 435.

9

The unsustainable aesthetics of sustainability: the sense of an ending in Jeanette Winterson's *The Stone Gods*

Adeline Johns-Putra

Jeanette Winterson's 2007 novel, *The Stone Gods*, is a critique of progress, both in the general sense of movement, journeying, or going forward, and in the specialised sense of human development, particularly the privileging of economic and scientific improvement that is often called the myth or narrative of progress. In the spirit of so many of Winterson's novels, *The Stone Gods* places its several protagonists on journeys, most often on journeys of self-discovery, in which a better understanding of self is the reward for a courageous opening up to new worlds and new experiences, new beings and new ways of being. Throughout the novel, such journeys of openness contrast with attitudes of certainty and even mastery over others. Somewhat perversely, the latter position of closure is associated with the myth of progress, so that progress is construed as not really progress at all but a retrograde process ending in destruction. More perversely still, the novel suggests that such destruction is doomed to be reiterated, so that the regress that is 'progress' is also on a constant loop of repetition. The book is, in the author's words, about our 'endlessly making the same mistakes' (Andermahr 2009: 131).

What this amounts to in the novel is a juxtaposition of two world-views, the first identified as having the potential to save the planet and the second as guaranteed to destroy it. On the one hand is an ethics of receptivity and embrace of human others, nonhuman others and even the biosphere at large; on the other is the impulse towards individual privilege through technological mastery of the world's finite resources, and that mastery measured in economic growth. The novel therefore replays some contemporary debates around sustainability – the idea that only an environmentally aware, empathetic re-orientation (of the kind associated with countercultural lifestyles and sometimes labelled deep ecology) can save the planet is set against the unsustainable practices of global capitalism, consumerism and individualism. However, between the two positions – which one could sketch as true sustainability as opposed to unsustainability – resides the compromise stance that is sustainable development, an attitude that is pivotal to understanding the novel.

Crucially, the novel's radical ethics of sustainability is a re-orientation not just of humans' attitude to the biosphere and its nonhuman species but of human modes of desiring and communicating with human and nonhuman others. It is a deliberate conflation of social and sexual norms with literary form and, as such, it is both an ethics and aesthetics of sustainability. In the novel, a radical openness to others is inseparable from a radical openness of narrative; psychological and physical journeys to the new and unknown involve the reader in false endings, circular stories and temporal tricks. Of all these, it is the novel's seeming desire not to end – in an ultimate gesture of openness – that proves most ambitious and therefore problematic. Winterson's rewriting of politics, ethics and aesthetics in the name of sustainability destabilises the very category of novel. I argue in this chapter that, thanks to its ethics and aesthetics of sustainability, *The Stone Gods* is predicated on its own unsustainability.

Before discussing Winterson's novel, I provide a brief overview of debates around sustainability in order to show how these are replayed in the novel's contrasting modes of sustainability and unsustainability. I then move from some basic assumptions behind models of sustainability to a rehearsal of theories of the novel, particularly to studies of endings, which are so relevant to an analysis of Winterson's alignment of sustainability with sexual and with narrative impulses, and particularly with romantic and narrative foreclosure. In analysing Winterson's novel, I suggest that the result – the novel's case for its own unsustainability – is borne out by its dénouement.

Sustainability, plural

A large part of the problem with defining sustainability is that it is often put to uses for which it was not intended, namely, uses beyond questions of economic utility and corporate accountability. Its short name of 'sustainability' means that one often forgets that its introduction into the mainstream of public thinking was as 'sustainable development'. When the United Nations World Commission on the Environment and Development, led by Gro Harlem Brundtland, published its report, *Our Common Future*, in 1987, it defined sustainable development, famously, as 'development that meets the needs of the present without compromising the ability of future generations to meet their own needs' (World Commission 1987: 43). As Simon Dresner points out in his discussion of Brundtland, sustainability was 'conceived as an attempt to bring environmentalist ideas into the central area of policy, which in the modern world is economics. It was to be the ground on which the mainstream was to consider the environmentalist case' (2008: 69). Taking account of this historiography of sustainability allows one to unknot the seeming paradox of sustainability: that it appears to be an exhortation simultaneously to preserve (our technological comforts, our economic development and ourselves as the human species) and to transform (our behaviour, our expectations, our destructive tendencies). Indeed, such a view allows one also to take a more understanding – if not entirely sympathetic – retrospective view of what Brundtland was attempting to achieve. The Brundtland Commission, along with the study of environmental economics that it in part inspired, values the environment in economic terms, as a 'good' or 'service'. So, for the environmental economist David Pearce, sustainable development may be more accurately termed 'sustainable utility' (Pearce et al. 1989: 32) and may be defined as 'non-depletion of capital', where capital includes not just human but 'natural capital' (Dresner 2008: 3). And sustainability's so-called 'three pillars' of economy, society, and environment have been translated by corporate accountants into the 'triple bottom line' (Elkington 1997: 2), where success is judged by offsetting and optimising performance in all three areas rather than treating social and environmental justice as absolute goods.

At the same time, one could say that the invitation to read sustainability as radical change and recalibration is buried within the Brundtland Commission report. The Commission's conclusion, reached after four years of collating and synthesising expert and public opinion, was that 'a new development path was required, one that sustained human progress not just in a few places for a few years, but for the entire planet into the distant future' (World Commission 1987: 4). In short, it is worth remembering that the Commission's ethical stance, though not its economic methodologies, was new.

Significantly, the new ethics of sustainability implied by Brundtland has increasingly been contrasted with – and discussed in isolation from – its economics. For environmental economist Timothy O'Riordan, sustainability 'in the conscience' and sustainability 'in the account book' (1993: 40) are very different things. More recently, Christian U. Becker has argued that the ethics of sustainability must be treated as distinct from scientific and economic understandings of the term. The latter two are identifiable as conservative, static modes, while sustainability ethics implies – and indeed necessitates – behavioural and psychological change (Becker 2012: 14). For Becker, such 'meta-structures' of science, technology and economics with their different but comparable emphases on individuality and utility can – and should – be challenged by an ethics of receptiveness and attentiveness to human and nonhuman others. Certainly, such a challenge lies at the bottom of attacks on Pearce for 'putting a price on the planet' (Dresner 2008: 112) and similar anxieties around terms such as 'ecosystem services' and 'carbon trading'. Openness, responsiveness and an acknowledgement of interdependence all mark an emerging normative ethics of sustainability, one set against both unsustainable consumerist practices and the perceived 'business-as-usual' compromises of sustainable development.

Narrative

Such a normative ethics, when viewed as an all-encompassing mode of not just being and doing but thinking and, indeed, meaning, has implications for literature and literary forms such as the novel. This alternative ethics of sustainability is recognisable in Winterson's novel as both an idealised basis for change *and* as a set of deliberately formal characteristics. The novel links an 'open' worldview (receptiveness to new paradigms beyond the status quo, to other beings beyond the familiar) with 'open' narrative form (non-linearity, repetition, and open-endedness). For one thing, it represents openness in terms of plot and practises openness in terms of structure; for another, it makes explicit comments throughout on the illusory nature of conventional narrative continuity and closure.

The yoking of narrative closure with social convention is not new, of course. Literary scholarship abounds with analyses that sometimes celebrate and sometimes critique our desire for plot resolution. The germinal text here is Frank Kermode's *Sense of an Ending*, which first appeared in 1967. According to Kermode, we render human existence and time significant by suggesting to ourselves that it is how everything ends (everything being existence, time, and so on) that gives it meaning. Specifically, Kermode distinguishes between two attitudes to time – *chronos* and *kairos*, where '*chronos* is "passing time" or "waiting time"' and '*kairos* is the season, a point in time filled with significance, charged with a meaning derived from its relation to the end' (2000: 47).[1] We prefer, says Kermode, kairotic preoccupations with meaning over the merely chronological experience of life – indeed, such kairotic concerns make the mundanity of the chronological bearable. It is this preference that drives the production, communication and reception of narrative. Narrative, in Kermode's formulation, is simply the space between beginning and end; as he puts it somewhat aphoristically, though we know that 'tick' must be followed by 'tock', the gap between them must be filled, even as the importance of that gap is determined solely by tick's leading to tock. Kermode's analysis highlights the psychological impulse that fuels the reader's desire for the end. The result is an important recognition, which Peter Brooks would go on to develop in his psychoanalytical studies of the 'masterplot' (1994), that the desire for deferral, with its prolongation of enjoyment, means nothing without the desire for dénouement, with its resolution and hence satisfaction.

It is a short step from this to psychoanalytical feminist and queer critiques of Kermode and, later, of Brooks. For example, second-wave feminist critics tended to identify Kermode and, to a lesser extent, Brooks as masterplotters themselves, perpetrators of a commitment to climax and closure that happens to suit the psychosexual inclinations of heterosexual male authors and audiences. In a not entirely nuanced attack, commentators such as Nancy Miller (1988) and Rachel Blau DuPlessis (1980) suggest that connubial happy endings are really a way of fantasising the limitation of female agency, and Susan Winnett (1990) famously condemns narrative climax as a patriarchal version of the pleasure principle. Most famous (or infamous) of all – and, it must be said, more pertinent to this analysis – is Lee Edelman's queer alignment of readerly and romantic impulses, which shows that the desire for endings – what in Kermode's terms is our kairotic desire for narrative 'concordance' – is part of a heteronormative sexual desire for dynastic stability and continuity. Edelman describes and castigates happy endings not as moments of closure *per se* but as fulfilments of nuclear family fantasies; for him, the trajectories of mainstream literature and film are governed by 'reproductive futurism' (2004: 2) – that is, the prevalent heteronormative and neoliberal logic of progress, procreation and posterity. The child, according to Edelman, is 'the perpetual horizon of every acknowledged politics, the fantasmatic beneficiary of every political intervention' (2004: 3). In the terms of Edelman's analysis, the conventional Western narrative of progress ends – or indeed never quite ends – in a vision of the future. This is a future governed by heteronormative reproduction with no place for sexual enjoyment *qua* enjoyment (Edelman uses the word '*jouissance*'), and thus with no place for queer, hybrid and other sexualities.

Thus, for Edelman, our kairotic desire for ending (inasmuch as, says Kermode, our endings give meaning to our beginnings and middles) is also a desire for reproductive continuity. To read Edelman alongside Kermode is to recognise that what we ask of our stories and of our lives are endings (for which, read *versions of the future*) that look like us. This strategy of plot resolution as posterity begins to sound suspiciously like the 'sustainable development' version of sustainability, the idea that we can – indeed, must – keep things going as they have been going. Contrasted with this, as I have already suggested, is narrative, ethical and environmental openness. Such a set of alignments is what Winterson attempts to effect in her novel. It is also what Karen Pinkus performs in her cultural critique of sustainability, in which she maps Edelman's analysis of reproductive futurism onto conservative forms of sustainability and then onto conventions of narrative closure. Pinkus focuses first on sustainability in 'terms of business,' where 'the future is actually calculated as part of a strategy of control'; for Pinkus, 'the "future" implied by sustainability [is] a future always already predetermined through strategic planning and regulation' (2011: 71). According to Pinkus, such a version of sustainability is, after Edelman, 'deeply imbricated in reproductive futurity' (2011: 73). Pinkus, however, also imagines and advocates a radical and open ethics of sustainability; in so doing, she shifts the terms of her discussion to narrative and discourse. She acknowledges that it 'is all too easy to dismiss sustainability as a misguided, liberal discourse that either forecloses the possibility of radical change or narrativizes in order to enfeeble the explosiveness of the momentary' and that, instead, we must 'take seriously the pleasure it contains *in potentia*' (Pinkus 2011: 74). Pinkus contrasts an economic version of sustainability that 'writes or implies a narrative coherence' (2011: 74) with a radical sustainability that enables 'risk', 'irony' and '*jouissance*' (74). Edelman, it must be said, imagines queerness as 'no future', as almost nihilistic closure; in contrast, Pinkus, by invoking Edelman but also celebrating the pleasure of potentiality, hints at a queer future of openness, both ethically and aesthetically speaking. Yet, Pinkus does not follow through on the contradictions raised by her analysis – contradictions about what this combined ontological and discursive openness might actually look like, particularly at the moment of narrative ending. As I hope to show, it is just this promise of openness, and just this problem of ending, with which Winterson's novel struggles and never quite comes to terms.

The Stone Gods: queer, quantum aesthetics

The Stone Gods is her tenth novel for adults (eleventh, if one considers her early graphic novel, *Boating for Beginners* (1985)), part of a considerable *œuvre* that began with the critical and popular success of *Oranges Are Not the Only Fruit* (1985). She has long been categorised in 'the boxes labelled "lesbian fiction" or "postmodern fiction"' (Pykett 1998: 53), though such labels are problematic. While novels such as *Oranges, The Passion* (1987) and *Sexing the Cherry* (1989) established her as a lesbian writer, Winterson herself has eschewed that label (see Morrison (2006) for an account of the vexed history of queer readings of Winterson). One could say more generally, as many critics do (for example, Onega 2006: 8; Andermahr 2009: 25), that her work is dominated by the idea of love, particularly, the theme of love as a power that transcends the real and the use of 'the metaphor of lovemaking as writing' (Andermahr 2009: 26). And, while her novels' tendency for experimental points of view, non-linear temporality and rich, repetitive style seems ample evidence of her postmodernist credentials, Sonya Andermahr, citing Susana Onega, rightly identifies a tension in Winterson's postmodern aesthetics: 'she disparages realism with its focus on narrative storytelling, yet extols storytelling as a human need and aesthetic principle' (Andermahr 2009: 27). Such a contradiction is at the heart of *The Stone Gods*, which privileges open-endedness as somehow true to an ethics of openness and receptiveness, while seeking closure and resolution in order to 'make sense' (to invoke Kermode's terms) of love and life.

The Stone Gods is better described as a set of 'novellas-in-a-novel' (Ellam 2010: 220). It is a collection in four chapters of three stories that take place, respectively, in the distant future, the eighteenth century and the near future; these revolve around three protagonists each named Billie Crusoe (or Billy, where the protagonist is male in the second story). In each narrative, Billie/Billy's psychological development is troped as a journey in which new worlds and new ways are discovered. As Andermahr reminds us, 'Winterson often places the theme of love in the context of a quest narrative' (2009: 26). In *The Stone Gods*, such journeys also enable love to become what is repeatedly described as 'an intervention' (Winterson 2007: 68, 183, 205)[2] into old, unsustainable ways. Each journey, then, is also a chance to short-circuit humanity's destructive tendencies with a new ethos, recognisable as an open ethos of sustainability.

In all three stories, Billie/Billy makes a strong connection with a character called Spike (or Spikkers, in the second story) and witnesses unprecedented environmental destruction. In the first chapter, Billie and Spike are part of a reconnaissance crew that leaves the dying planet Orbus to establish the habitability of the newly discovered Planet Blue. Orbus is a version of a future Earth (although, by the end of the novel, it becomes apparent that Planet Blue is really an ancient Earth). The Billie of this first chapter is a scientist who despises her civilisation's Central Power for encouraging its citizens to stay in thrall to cosmetic surgery, celebrity culture and artificial, climate-controlled environments, while it competes with the planet's other superpowers – the Sino-Mosco Pact and the Caliphate – to use up the last of the planet's resources. Spike is a type of advanced humanoid robot – a Robo *Sapiens* – with evolutionary capabilities. Billy in the second chapter is a sailor on the second voyage of Captain Cook's *Resolution*, accidentally left behind on Easter Island, where he falls in love with a Dutchman, Spikkers, and – evoking the cultural commonplace that the Easter Islanders sacrificed the viability of their ecosystem to their religious beliefs – witnesses the destruction of the island's last tree in a power struggle. The Billie of the third and fourth chapters lives on Earth in a near-future time known as 'Post-3 War' (158), the aftermath of a devastating world war amidst the ruins of a climate-changed planet. Her job involves educating Spike, this time a cyborg created to provide objective, rational government of the country. Britain is run by a faceless corporation called MORE, which has rebuilt the war-torn economy and now asserts complete control over it. But MORE is also the name of the corporation behind the Central Power of the first chapter. Thus, Planet Blue is Planet Earth, and is committing identical, not merely similar, mistakes to those made by humans on Orbus. The incidental story of Planet White, on which 'humans, or whatever they were, massively miscalculated and pumped so much CO_2 into the air that they caused irreversible warming' (68), is a reminder to the reader that the destruction of the biosphere will be reiterated as many times as there are biospheres.

The novel makes explicit the similarities between loving planets and loving people. Hearing of Planet White, Billie bemoans how humans 'keep making the same mistakes over and over again,' to which Pink – a typically youth-obsessed, cosmetically enhanced citizen of Orbus – responds that '[w]omen are just planets that attract the wrong species' (69). Certainly, the unloved Orbus presents as the victim of an abusive relationship. Asks Billie, sarcastically, 'We didn't do anything, did we? Just fucked it to death and kicked it when it wouldn't get up' (8). Even as it is on the receiving end of exploitation, it is also the setting for it, particularly of a shallow kind of masculinist exploitation – what one critic has described as a 'mutated form' of 'patriarchal gender dichotomies' (McCulloch 2012: 65). On Orbus, men no longer need women for procreation, women resort to identikit artificial beauty to achieve desirability, and men turn to young girls for 'something different when everything has become the same' (Winterson 2007: 21). Planet Blue proves this maxim. Advertised on Orbus with lines from Donne's poetry ('She is all states, all Princes I'), it is construed as feminine and ripe for conquest. Those lines, we learn, have come from Captain Handsome, the man who discovers Planet Blue and, later, engineers what he calls 'species-control' (82), an asteroid collision to kill the planet's dinosaurs and to create conditions in which humans can live.

Planet Blue, however, turns out to evoke something other than masculinist domination; it becomes – and then stands in for – the object of a queer desire. Billie's journey to Planet Blue coincides with the awakening of her desire for cyborg Spike. What is significant for Billie is not that she is falling in love with a woman (indeed, Billie's sexual orientation prior to joining the reconnaissance mission is never identified) but that she is falling in love with a robot: 'My lover is made of a meta-material, a polymer tough as metal but pliable and flexible and capable of heating and cooling, just like human skin' (83). Abigail Rine's reading of the novel as queering the human, not merely the heteronormative, is pertinent here: 'Billie is a queer heroine, not because she is a lesbian (though she is), but because she transgresses fundamental norms. ... Queerness, for Winterson, is not simply non-heterosexuality, but that which intentionally challenges and exceeds the constraints of the normal' (2011: 77).

Billie's first sight of the unknown planet glosses the radically new experience opened up by Spike: 'I can't wipe out the yes. One word and a million million worlds close. One word, and for a while there's a planet in front of me, and I can live there' (Winterson 2007: 83–4). Billie's lyrical description of the planet – 'But there she is, sun-warmed, rain-cooled, moon-worshipped, flanked by the stars. There she is. Planet Blue' (84) – thus provides an alternative version to Donne's lines adopted as advance publicity for the planet. In pointed contrast, Spike is desired by Handsome in precisely the masculinist terms of Donne's poetry; indeed, it is his discovery of those lines and use of them to describe Spike that subsequently informs the marketing of Planet Blue to Orbus.

Spike and hence Planet Blue are identified as sites of resistance not to masculinism as such but to heteronormativity and reductivism together, in a neat reproduction of Donna Haraway's terms of the cyborg as a feminist category. For Haraway, the cyborg privileges multiplicity of identity and rejects essentialism, and gestures to 'lived social and bodily realities in which people are not afraid of their joint kinship with animals and machines, not afraid of permanently partial identities and contradictory standpoints' (1991: 154). In contrast, both Orbus and its over-sexed men and women are being destroyed by a selfish anthropocentrism: 'Humans always assumed that theirs was the only kind that mattered. That's how you destroyed your planet' (Winterson 2007: 79), says Spike. 'But,' she continues, 'you have a second chance. Maybe this time …' (80). Spike and Planet Blue represent that second chance for Billie and for us: 'I looked at Spike, unknown, uncharted, different in every way from me, another life-form, another planet, another chance' (90). The opportunity to assume 'partial identities' and 'joint kinship' with others is exemplified by Spike's ability to evolve into something that unites the best of human and robot: 'We are solar-powered and self-repairing. We are intelligent and non-aggressive' (79). Little wonder that the vision she has for Planet Blue is 'to develop a hi-tech, low-impact society, making the best of our mistakes here, and beginning again differently' (39). Billie, similarly, must learn to embrace both machine and animal others (quite literally, as she ends her life on Planet Blue in the company of Spike and Three Horn, the 'hog-hippo hybrid' she conveniently adopts as a pet). The embrace of others, of possibility, of difference is troped as a radical new sustainable ethics for the future. Spike explains their first kiss in the context of a 'quantum universe' that 'is potential at every second' (75).

This quantum, cyborgian ethics with its tone of openness and unpredictability has profound implications for the chapter's conclusion – its moment of closure. Billie and Spike die soon after arriving on Planet Blue, victims of the asteroid impact facilitated by Handsome; yet they are not, at the same time, victims of masculinist domination. As death approaches, humanity and technology meet halfway. Billie and Spike make love, which Billie experiences as a journey of enlightenment: 'She is the missing map. She is the place that I am' (107); meanwhile, Spike develops a heartbeat.

Death brings love; love is, for Billie, 'a journey on foot to another place' (109) and, for Spike, 'the chance to be human' (110). The novel insists, almost mantra-like, on birth instead of death, dream instead of sleep, and beginnings instead of ends: 'Things dying ... things new-born' (112) and 'Close your eyes and sleep. Close your eyes and dream. This is one story. There will be another' (113). Closure is resisted, renaissance anticipated.

Indeed, another story follows on, signalling not the fixed, reproductive futurity of 'happily ever after' as critiqued by Edelman, but another journey outward, another opportunity to witness and resist regressive 'progress'. Billy the sailor, the protagonist of the next chapter, understands the self as a ship and the soul as 'the seabird that ploughs the wake of a Ship and then flies away no man knows where' (Winterson 2007: 131–2). This chapter juxtaposes once more the journey to and discovery of love by an enlightened pair with the destruction of ecosystems by ignorant others. Again, the protagonist Billy finds love with someone who is at once a hybrid figure and a protector of the environment: the half-Dutch, half-native Spikkers, a man intent on saving the ravaged island. The events on the island microcosmically mirror the failures on planet Orbus. The devastation of Easter Island is explained as the result, in the first instance, of the use of wood to transport and erect the giant *moai* under the Ariki Mau, or White Man, and his regime of ancestor worship. So much accords with accepted (though now contested) versions of Easter Island's history.[3] In Winterson's narrative, however, devastation is exacerbated by the Ariki Mau's rival, the Bird Man, who uses up the island's dwindling resources for himself and his followers. Thus, unthinking environmental exploitation, followed by corruption and resource war, lay waste to the island. These blatantly unsustainable practices prompt Billy to ask, quoting the Bible, 'What should it profit a Man that he gain the whole World and lose his own Soul?' (131). Whatever the religious impulses of the Ariki Mau, the island's deforestation is interpreted by Billy – in an echo of Orbus – as having more to do with material greed than spiritual fulfilment.

Both Billies' journeys of queer discovery coincide with love for another and love for the nonhuman environment. While the first Spike opens up for Billie a 'hi tech, low impact' future of possibility, Spikkers unfolds for Billy the island's 'ghastly history' (132) and then reveals his plan to replace the Bird Man and to restore both civil and ecological stability. Moreover, just as the first Billie discovers her love for Spike as they approach a new world and a newly possible worldview, Billy realises he is in love with Spikkers in the midst of Spikkers's attempt to win the race to be the new Bird Man.

Spikkers's death, however, hints at another kind of journey, one that – I will go on to argue – might be aligned with the essentially preservative and conservative impulse of 'business-as-usual' sustainable development. So far in the novel, the contrast exists between repetition (the endless loop of narratives of progress and unsustainability) and possible 'intervention' (a short-circuiting of the loop by a new, open, queer/cyborgian ethics of sustainability). Spikkers introduces to Billy the desire for home, a desire that is both a reaching-out and a turning-back, figured by the openness of outer space and the closure of return to his father's homeland. In a gesture that combines both impulses and is overlaid by a further desire for Billy, 'Spikkers pointed up to a bright and steady star close to the moon. ... "Holland," he said, kissing my fingers, one by one by one, and until my hand became a five-pointed star' (129). This confuses Billy's ideas of home – is Plymouth 'nearer than a Holland star – or easier to believe?' (130), he wonders. He comforts the dying Spikkers with the thought of home, represented by the tall house on his Delft tile, but also signified by the night sky: 'In the sky there is a star called Holland and the tall wooden houses of Amsterdam are clear to be seen' (139). And this is Billy's journey too, for he looks forward, after Spikkers's death, to being reunited with him in such a house – or, rather, to their souls being reunited. The 'white Bird [that] opens its wings' (140) in the chapter's final line becomes a reference simultaneously to Spikkers's soul on its voyage of return and to Billy's soul in anticipation of his reunion with Spikkers. The chapter ends, then, on a note of openness, but this openness looks forward, paradoxically, to the closure that comes with return and reunion.

In both chapters, ethical, environmental openness and stylistic openness are one and the same thing. New loves and new worlds are also new stories. The 'intervention' that Spike names is not just a political or psychological recalibration; it refers to the rewriting of narratives of progress and mastery. Indeed, it refers to the rewriting of narrative as we know it. Intervention offers itself as a key term, suggesting an interruption of linear trajectories and expectations, and the novel's rich lyricism (what Jean-Michel Ganteau has perceptively labelled Winterson's 'baroque aesthetics') references an ability, in Ganteau's words, to 'overflow the frame and all possible margins' (2005: 193). This conjunction of formal with ethical openness is also noted by Rine's analysis of the novel:

> Winterson's love intervention [is] twofold: first, she presents love as a renewed form of relationality that is not constrained by the dominant order, one that seeks mutuality and intimacy rather than appropriation and objectification; second, Winterson also suggests the possibility of a love between reader and text that opens new worlds, new potentialities. (Rine 2011: 79)

The suggestion occurs, further, that such formal interruptions and deviations are more true to life and to the universe (which is, after all, a 'quantum' one, according to Spike). Billie looks back on her life and remembers not 'the stories with a beginning, a middle and an end, but the stories that began again, the ones that twisted away, like a bend in the road'. *Pace* Kermode, she decides that '[t]rue stories are the ones that lie open at the border, allowing a crossing, a further frontier ... Like the universe, there is no end' (Winterson 2007: 106).

Each new tale in the novel, of course, performs this idea self-referentially, suggesting that the previous story is, somehow, still open. Rine reads this, after José Esteban Muñoz, as a kind of 'queer utopianism' (Rine 2011: 70); Muñoz (2006), like Pinkus, offers a positive inflection of Edelman's negative identification of queerness as a rejection of the future. Nonetheless, each new story also, problematically, confirms ideas of closure, certainty and progress in reminding us that we are doomed to repeat our mistakes. This, as the reader moves on to the final two chapters, produces a creative tension. Thus far, the first story has opened up into a new one, but that new one has merely confirmed our propensity for repetition. With the third chapter and another beginning, the attentive reader may well wonder if an intervention into the cycle of destruction might be made at last. What that reader encounters, however, is the ethics and aesthetics of return.

Closure, return and repetition

This final tale repeats the theme of the retrograde myth of progress, but also dashes any expectations the reader might have (honed over the previous two tales) of queer, open love – and, with it, of a radical, open ethics of sustainability. In London, a woman named Billie discovers a manuscript entitled 'The Stone Gods' on the Tube, making manifest the novel's assumptions, expressed by the first Billie, that another and another story will be told. This third Billie, however, is defined not as a journeyer but as a castaway. Here, Billie's surname of Crusoe finally comes into its own: she is 'shipwrecked on the shore of humankind' (Winterson 2007: 148). Love, in the case of this Billie, is defined by loss, the loss of her mother who was forced to abandon her as a baby and for whom she obsessively seeks. She will, she says, 'never stop looking', and lives with the thought of her mother as though 'in an echo of another life' (149). She thus calls to mind less the impulse of queer love as signified by the previous incarnations of Billie and Spike and more the nostalgic desire of Spikkers for his father's home and of Billy for the dead Spikkers. The longing for familial return signified by the 'Holland star' in the previous story – and thus what Edelman terms the 'pronatalism' (2004: 17) of reproductive futurism – becomes the dominant note of this story.

Accordingly, the Spike of this vignette is hardly Billie's soulmate. Indeed, she could be regarded as inferior to the first Spike both physically and ethically. This Robo *Sapiens* is nothing more than a robotic head, developed to provide advice to the MORE corporation: 'She has no body because she won't need one ... to take the planet-sized decisions that human beings are so bad at' (Winterson 2007: 158–9); these are 'neutral, objective decisions ... for the global good' (198). Attuned to MORE's corporate goals, this Spike believes in the possibility of 'the transition from the economics of greed to the economics of purpose' and insists that 'economics of purpose is not about making money: it is about realigning resources' (164). The 'economics of purpose', part of MORE's 'new world order' of 'modest and eco-conscious members' (165), comes suspiciously to sound like a regime of sustainable development, invested in environmental and social justice as ends but also in economic efficiency and resource management as means to those ends. No surprise, then, that Billie is aware that '[n]either art nor love fits well into the economics of purpose, any more than they fitted into the economics of greed. Any more than they fit into economics at all' (169). This Spike does not offer a queer, quantum ethics of love as intervention. Notably, Billie's list of '[w]hat it means to be human' includes not just 'art' and 'love' but raising children who 'know the value of the world and not its economic potential' and 'all the invisibles never counted by the GDP' (198). In the terms of this chapter, humans equal art and love; robots equal the economics of both purpose and greed. There is precious little room on Billie's list for journeys of cyborgian discovery or, put another way, queer utopianism.

The story's ethical faultline – capitalism on the one hand and 'what it means to be human' on the other – is also the boundary between Tech City and Wreck City. Tech City is the MORE-controlled zone where Billie and Spike live, and Wreck City is a 'No Zone – no insurance, no assistance, no welfare, no police' (179). If Tech City represents the economics of purpose, Wreck City offers something like openness, but not necessarily love. Wreck City's black market economy may have been branded 'Bad Capitalism' by MORE but, on Billie's closer inspection, she realises that it operates on a pre-capitalistic system of bartering. Its town centre – the Playa – is a carnivalesque space that serves as meeting place, market square, performing ring, 'fairground, bacchanal, dream' (224). Home to 'twenty alternative communities' (207), Wreck City is where Spike decides to join a group of lesbian vegans called Chic X, so-called because they regret the damage done by the asteroid at Chicxulub – they regret, in the novel's twisted timeline, Handsome's anthropocentric geoengineering. Spike's defection from the economics of purpose to an alternative ethics is marked by her sexual encounter with one of the Chic Xs, but this is a comic version of the cyborgian epiphany experienced by the first Spike and Billie. This is not embraced by Billie, who dismisses it sarcastically: 'Great. The robot that was designed to become the world-sage has had oral sex with a teenager called Nebraska and become a drop-out free-love silicon guru' (210). The barman aptly named Friday who serves as Billie's guide in Wreck City does little better when it comes to offering love. He reintroduces her to many things long gone from Tech City, such as books and wood, and provides her with information that allows her to recalibrate her understanding of Post 3-War socio-economics; in short, like Crusoe's Friday, he provides the 'shipwrecked' Billie with company and assistance on her island. Yet, unlike Spikkers, he does not provide love. That the moment of romantic tension between them – 'He looked at me. I nearly touched him' (199) – is unresolved is a signal that what Billie longs for will not come from Friday. A classic 'happily ever after' is not enough to rescue this castaway from the wreckage of neoliberal unsustainability. But neither can it come from the newly converted Spike, whose queer, open ethics of sustainability has been irretrievably lampooned.

This Billie wants a landing-place rather than a journey; her goal is one of reversion rather than openness and possibility. Her default state, she recognises, is loneliness, and the 'opposite of loneliness isn't company, it's return. A place to return' (175). Then, later, she realises that the landing-place she seeks 'isn't a place at all: it's a person, it's you' (200). In the first two stories, the free-floating second-person pronoun seems to represent the ultimate and unknown object of desire, before it is gradually revealed to be Spike or Spikkers. In this story, 'you' is the first object of desire, the mother. At the end of the story, Billie is, indeed, reunited with this original love, but in death. Shot by MORE soldiers, she experiences a phantasmic out-of-body sensation of walking to the stone farmhouse her mother used to walk to when pregnant with her and, at the gate to the farmhouse, is met by her dead mother. Just before she dies, Billie repeats the words of the first Billie and Spike:

> A quantum universe – neither random nor determined. A universe of potentialities, waiting for an intervention to affect the outcome.
>
> Love is an intervention.
>
> Why do we not choose it? (Winterson 2007: 244)

This repetition forces on the reader the suggestion that the return to maternal love is as much an intervention as is the journey to openness and cyborgian love. Certainly, the story of Spikkers's death and Billy's longing for death had suggested that death might be a way to re-experience and re-encounter love. Nonetheless, it is difficult to read in the third Billie's death anything like openness, for it speaks primarily of return and regress, of endings and retracings. Most of all, it bespeaks a mode of fixed continuity – indeed, it betokens a reproductive futurism.

This insistence on return represents a particularly problematic 'intervention' for the novel, given its promise of open and radical aesthetics. Some critics have explained the contradiction – one might even say anti-climax – that Billie's death and filial reversion make of the dominant motif of rebirth, openness and potential, by insisting that maternal love and queer love are equivalent and that both imply possibility. For Rine, the Billies all proceed on an arc of love and desire; it is simply that desire, for the final Billie, is initially motivated by loss. Despite seeming evidence to the contrary, Rine finds that the novel 'leaves open the possibility of a queer future that is not merely lethal repetition and affirms an idea of nonheteronormative reproduction through the creative and transformative potential of language' (2011: 83). Similarly, Fiona McCulloch insists on both the queer and feminist possibilities at play throughout, stating that each Billie is capable of 'teaching a different story to phallocratic grand narratives that offer[s] a moral compass of queer cosmofeminism' (2012: 67–8). McCulloch even implies that the final Billie's life-in-death experience is yet another example of 'the capacity of human love that transcends divisions' (2012: 72). Onega, similarly, proposes that Billie's death is overturned by a reunion with the mother that is 'pregnant with possibilities' (2011: 297); only a 'realist reader', suggests Onega, would have trouble with such an ending. Only Julie Ellam protests that the novel's maternalist ethics are incompatible with cyborgian ones, for 'the lost mother as a symbol for the physical disintegration of the planet becomes restricted in its power' (2010: 224) in the light of the novel's earlier queer claims. Unlike Onega's putative 'realist reader', I am not suggesting that a fictional death is always a moment of closure. What I am concerned to point towards, however, is how the last Billie's regression into the past cancels out the journeys imagined by the previous Billies and promised by their beloved Spikes. This Billie insists not so much on ending (there is no 'happily ever after' for her, as we have seen) but on returning to beginnings. She thus embodies the kind of conservative and conservationist impulse that, much to Edelman's regret, makes our lives and our stories meaningful.

For this return is necessarily a return to the first Billie – the farm to which the third Billie returns is identifiable (with its apple tree, water barrel and gate) as hers. Prior to this, the third Billie had been emitting a sense of filial yearning her whole life, 'calling you, across time. Steadily sending the signal, sure that, one day, you will hear' (Winterson 2007: 220). Sitting in the disused giant dish of the Lovell telescope, the third Spike picks up 'a repeating code bouncing off the surface of the moon' (237); this, the signal sent from the first Billie and Spike to their distant future counterparts, is the longed-for maternal response. Even the copy of 'The Stone Gods' left on the Circle Line offers itself for interpretation as a gift from the first Billie as mother to the third Billie as daughter: when Spike asks who left it there, Billie quixotically says, 'It was me' (241). The novel, it would seem, is literalising its earlier promise that '[t]rue stories are the ones that lie open at the border, allowing a crossing' (106). But this crossing is now not a crossing into the unknown in the spirit of potentiality; it is a crossing back in the spirit of return. The journey structured here is one of a closed loop, rather than an opening up. It is as if the centrifugal motion of the journey of discovery has been forcibly reversed into a centripetal turning-within.

It is not simply that this Billie insists on a return to the first Billie as mother; it is also that such a return structures parent–child relationships as closed circuits. This echoes Edelman's critique of reproductive futurism as an essentially preservative rather than provocative agenda. The reader should recall that the first Billie's examples of '[t]rue stories that lie open at the border' come from her own life, things left unfinished 'not because I left it too soon, but because it had a life of its own that continues without me' (106). This includes children, who 'I suppose, are always unfinished business: they begin as part of your own body, and continue as separate as another continent' (106). The first Billie's construction of the mother–child relationship is one of potentiality and possibility; it points to the life of the child as a life that is inevitably independent of its mother, as a new world, a new story. The novel's conclusion, in contrast, construes the child's life as eternally interlocked with the mother's, implying, by association, that stories are not just closed but in constant playback.

In my reading of *The Stone Gods*, the novel has inadvertently performed that which it sought to defy. Paradoxically, its attempt at openness, in the spirit of true environmental sustainability and in defiance of closure and certainty, results in a return to the beginning and therefore to something more problematic than closure: repetition. The novel strives hard throughout to associate repetition with a failure of imagination, intervention and love; yet, its nostalgia for love, figured as filial return, becomes an ultimate act of repetition.

If the novel insists on openness in terms of an ethics of radical sustainability, an erotics of cyborgian/queer utopianism, and an aesthetics of narrative experimentation, then what are the ramifications of its final embrace of reproduction, replication and repetition? In concluding her novel on this note, Winterson (perhaps unwittingly) discloses something important about human desire, namely its propensity for sustainability in its conservative rather than revolutionary sense, emanating from its profound dependence on the familiar. It is not simply that her novel – like all novels – must end; it is noteworthy that her novel – like all novels, as Kermode would have it – strives to provide the reader with a sense of kairotic significance, a view of beginnings and endings together, each meaningless without the other. Is it the case that, no matter how much Billie – and we – may reach out to others, we make sense of that reaching out by referring back to our origins? This is not simply another way of saying that we can only know how far we have gone by considering where we began; it is an admission – if a reluctant one – that no matter how much we embrace a future of otherness, we want something in that future to resemble ourselves. Rather than deny our latent preservative tendencies, we may do well to acknowledge them; rather than bemoan the compromises made in the name of sustainable development, it behoves us to ask why these were necessary. The key question for any honest appraisal of sustainability – radical, conservative or otherwise – may be not so much what we want the future to look like and what potential it may hold, but what part of ourselves we want to sustain.

Notes

1 Kermode borrows the terminology from theology, and further notes, 'We can use this kind of language to distinguish between what we *feel* is happening in a fiction when mere successiveness, which we *feel* to be a characteristic in the ordinary going-on of time, is purged by the establishment of a significant relation between the moment and a remote origin and end, a concord of past, present, and future' (2000: 50, author's italics).

2 All subsequent references to *The Stone Gods* are from this edition, unless otherwise stated.

3 The oft-repeated claim that the people of Easter Island, or Rapa Nui, devastated their forests to assist in building the *moai*, or stone gods, has been recently challenged: see Christian M. Stevenson et al. (2015)

References

Andermahr, Sonya 2009. *Jeanette Winterson*. Basingstoke: Palgrave.

Becker, Christian U. 2012. *Sustainability Ethics and Sustainability Research*. Dordrecht: Springer.

Brooks, Peter 1994. *Reading for the Plot: Design and Intention in Narrative*. Oxford: Clarendon Press.

Dresner, Simon 2008 [2002]. *The Principles of Sustainability*. London: Earthscan, 2nd edn.

DuPlessis, Rachel Blau 1980. *Writing before the Ending: Narrative Strategies of Twentieth-Century Women Writers*. Bloomington: Indiana University Press.

Edelman, Lee 2004. *No Future: Queer Theory and the Death Drive*. Durham, NC: Duke University Press.

Elkington, John 1997. *Cannibals with Forks: The Triple Bottom Line of 21st Century Business*. Oxford: Capstone.

Ellam, Julie 2010. *Love in Jeanette Winterson's Novels*. Amsterdam: Rodopi.

Ganteau, Jean-Michel 2005. '"Rise from the Ground like Feathered Mercury": Baroque Citations in the Fiction of Peter Ackroyd and Jeanette Winterson', *Symbolism: An International Journal of Critical Aesthetics* 5: 193–211.

Haraway, Donna 1991. 'A Cyborg Manifesto: Science, Technology, and Socialist-Feminism in the Late Twentieth Century'. In *Simians, Cyborgs and Women: The Reinvention of Nature*. London: Free Association Books, 149–81.

Kermode, Frank 2000 [1967]. *The Sense of an Ending: Studies in the Theory of Fiction*. Oxford: Oxford University Press.

McCulloch, Fiona 2012. *Cosmopolitanism in Contemporary British Fiction: Imagined Identities*. Basingstoke: Palgrave Macmillan.
Miller, Nancy K. 1988. *Subject to Change: Reading Feminist Writing*. New York: Columbia University Press.
Morrison, Jago 2006. '"Who Cares about Gender at a Time Like This?" Love, Sex and the Problem of Jeanette Winterson', *Journal of Gender Studies* 15: 169–80.
Muñoz, José Esteban 2006. 'Thinking beyond Antirelationality and Antiutopianism in Queer Critique', *PMLA* 121: 825–6.
Onega, Susana 2006. *Jeanette Winterson*. Manchester: Manchester University Press.
Onega, Susana 2011. 'The Trauma Paradigm and the Ethics of Affect in Jeanette Winterson's *The Stone Gods*'. In *Ethics and Trauma in Contemporary British Fiction*. Ed. Susana Onega and Jean-Michael Ganteau. Amsterdam: Rodopi, 265–98.
O'Riordan, Timothy 1993. 'The Politics of Sustainability'. In *Sustainable Environmental Economics and Management: Principles and Practice*. Ed. R. Kerry Turner. London: Belhaven Press, 37–69.
Pearce, David, Anil Markandya and Edward B. Barbier 1989. *Blueprint for a Green Economy*. London: Earthscan.
Pinkus, Karen 2011. 'The Risks of Sustainability'. In *Criticism, Crisis, and Contemporary Narrative: Textual Horizons in an Age of Global Risk*. Ed. Paul Crosthwaite. New York: Routledge, 62–77.
Pykett, Lynn 1998. 'A New Way with Words? Jeanette Winterson's Post-Modernism'. In *'I'm Telling You Stories': Jeanette Winterson and the Politics of Reading*. Ed. Helena Grice and Tim Woods. Amsterdam: Rodopi, 53–60.
Rine, Abigail 2011. 'Jeanette Winterson's Love Intervention: Rethinking the Future'. In *Sex, Gender and Time in Fiction and Culture*. Ed. Ben Davies and Jana Funke. Basingstoke: Palgrave, 70–85.
Stevenson, Christian M. et al. 2015. 'Variation in Rapa Nui (Easter Island) Land Use Indicates Production and Population Peaks Prior to European Contact', *Proceedings of the National Academy of Sciences* 112 (4): 1025–30.
Winnett, Susan 1990. 'Coming Unstrung: Women, Men, Narrative, and Principles of Pleasure', *PMLA* 105 (3): 505–18.
Winterson, Jeanette 2007. *The Stone Gods*. London: Penguin.
World Commission on Environment and Development 1987. *Our Common Future* (Brundtland Report). Oxford: Oxford University Press.

10

A modest proposal for a less natural lifestyle: the paradoxes of sustainability and Michel Houellebecq's *The Possibility of an Island*

Hannes Bergthaller

Maggots in a box, or *Our Common Future*

When I was about twelve or thirteen years of age, my family spent a long summer vacation at the holiday home of a relative in Spain. Upon our return, I went straight to the kitchen cupboard to fix myself a bowl of granola. I opened the tupperware container and gasped: it was filled to the brim with scaly, reddish-brown maggots. Not a writhing mass – as I remember them (and I am aware that my memory is probably playing tricks on me here) – most of the maggots were already dead, starved to death after having completely transmuted my granola into the pulpy substance of their own tiny bodies (all the while I had been disporting myself on sunny Mediterranean beaches); or, who knows, perhaps they had resorted to cannibalism and were now maggots to the second or third power, so to speak. This image etched itself indelibly into my memory. Even at the time, I connected it to the stories about human overpopulation, forest dieback and wholesale environmental destruction, which, in the mid-1980s, were everywhere in the German news media. To me, the maggots appeared as a terrifying image of humanity itself, as it was portrayed in these stories: a wildly proliferating mass, voraciously consuming whatever resources came into its path, terminally blind to the 'limits of growth'.

Just around this time, the UN's World Commission on Environment and Development, led by former Norwegian prime minister Gro Harlem Brundtland, was conducting public hearings and collecting expert testimony from all over the world to prepare a document that would be published in 1987 under the title *Our Common Future*, but became better known as the 'Brundtland Report'. It was this text which first introduced the phrase 'sustainable development' into our vocabulary and thus established the policy platform on which the 'Earth Summit' at Rio de Janeiro was convened in 1992. It also informed the whole raft of policy initiatives that were launched in the wake of this meeting, from Agenda 21 to the Kyoto Protocol. Organisations of many different stripes (e.g. governmental institutions, corporations and universities) were quick to embrace the principle, not only because it offered a convenient way of signalling to the public that they took environmental concerns seriously, but also because it neatly dovetailed with their own need to make internal processes more transparent and efficient. Especially in Europe, the ascendancy of sustainability coincided with the rise of ecological modernisation, a discourse which sought to de-politicise environmental issues and reframe them as problems that could best be solved through the application of scientific and economic expertise. Considered in this light, today's 'gospel of sustainability' (Emerich 2011) appears much like an updated version of the 'gospel of efficiency' that drove the conservation movement of the late nineteenth and early twentieth centuries (Alaimo 2012: 558–9).

As the evangelical moniker suggests, the concept also filtered down into the realm of personal belief and everyday practice, where it became amalgamated with the new forms of health-consciousness and reflexive consumerism which developed during the 1990s. This development is epitomised by the 'Lifestyle of Health and Sustainability' (LOHAS) label, an acronym popularised by social psychologists Paul H. Ray and Sherry Ruth Anderson with their book *The Cultural Creatives* (2000). There, it designates the consumer profile of the eponymous demographic: a cognitive elite of progressive, environmentally conscious people who, Ray and Anderson claimed, had rejected 'crass commercialism' (2000: 329) and were ushering in a profound social transformation through their virtuous consumer choices. In the world of LOHAS, sustainability is identified with the quest for individual health, spiritual wellbeing and a 'more natural lifestyle' (Emerich 2011: 137). Looking at LOHAS advertising and trade magazines, one begins to suspect that the true office of sustainability may have been to make deep ecology safe for consumer capitalism.

While these two versions of sustainability are in many ways quite distinct, both indicate just how deeply the concept has become ingrained in the political, economic and social status quo. They highlight how talk of sustainability has become a way of expressing our desire to change things so as to keep them pretty much the way they are. Understandably, scholars from the humanities who wish to take sustainability seriously therefore often see their task as a kind of a rescue mission: they wish to 'reclaim' the concept from the technocrats and the marketing experts (Keller 2012: 581) and to restore to it some of its critical edge. In this essay, I want to argue that such an effort must be based on the recognition that the paradoxes which trouble the discourse of sustainability are not merely symptoms of its co-optation; if anything, the opposite is the case: it is because of the deep ambiguities built into the concept that sustainability is so easily enlisted for spurious aims. Sustainability is a problem springing from the naturalness of human beings, from the fact that their survival as a species is conditioned by the same sorts of constraints as those which apply to all other living creatures. But for species survival to become a properly political issue, it must be qualified in some fashion. In order for it to be something more than applied biology, sustainability cannot just be about securing the existence of a viable population of *homo sapiens*; it must be concerned with sustaining a *particular* form of human life, a good life that would justify the effort of sustaining it, of staving off the fate of extinction awaiting all biological species. The only way in which we can know what a good life is, is through our own experience. (Of course, we are awash in stories of the good life, but these only make sense to the extent that they resonate with personal experience). For me, sustainability means that I always want there to be a boy playing on a Mediterranean beach. At its limit, sustainability abuts an impossible, narcissistic desire for immortality – a desire rudely negated when I find myself addressed merely as member of my own species.

Sustainability talk of the sort found in corporate mission statements and LOHAS brochures serves to obscure this hard aporetic knot; the task of the environmental humanities must be to keep it firmly in our view. Fiction is an indispensable ally in this effort because it can more readily dispense with polite compromise and scholarly circumspection, and engage in forms of hyperbole and imaginary amplification which help clarify what the actual stakes are. After discussing the troubled relationship between sustainability and neo-Malthusianism in the next section, I will turn to a text which does precisely this, namely Michel Houellebecq's *The Possibility of an Island*. The novel, I argue, can be read as an instance of satire in the Juvenalian mode of Swift's *A Modest Proposal*, which pushed the utilitarian logic underlying England's colonial policy in Ireland to a horrific extreme so as to expose its cruel hypocrisy. In a somewhat similar manner, *The Possibility of an Island* takes the confused desire for a more natural lifestyle that animates so much of the discourse on sustainability and extrapolates from it a future which reveals the latter's contradictory and potentially dehumanising logic. Sustainability, I conclude, is ill-suited for service as a first principle. If it is to play a role other than that of a weak anodyne for contemporary anxieties or a necropolitical calculus, it can only be that of a necessary stopgap for a problem that must remain essentially insoluble.

Sustainability, neo-Malthusianism, and the politics of emancipation

Humanist efforts to recuperate the concept of sustainability are often driven by the conviction that genuine sustainability can only be achieved from below, through the political mobilisation of civil society. Such efforts can refer themselves back to the original Brundtland Report which, with almost thirty years' hindsight, turns out to be not only a remarkably prescient, but also a surprisingly radical document. Environmental degradation and socio-economic inequality, the commission proposed, were problems which needed to be tackled together and which no nation could hope to solve on its own. At a time when environmentalists in most Western countries were still focused on the protection of a natural world imagined as standing apart from social concerns, the UN Commission insisted that 'the rights of people to adequate food, to sound housing, to safe water, to access to means of choosing the size of their families' (World Commission 1987: xi) should be conceived of as environmental issues. Failure to address them would not only harm human beings, but also undermine any effort to safeguard the natural environment. Furthermore, the Commission pointed out, arriving at a stable and just solution to these problems required 'political systems that secure effective citizen participation in decision making and ... greater democracy in international decision making' (World Commission 1987: 8). The report itself sought to illustrate what this would entail by including verbatim testimony from engaged citizens – for example, from a local organiser working to improve the living conditions of slum dwellers in Indonesia (World Commission 1987: 254). As it was laid out in *Our Common Future*, the notion of sustainable development erased customary boundaries between political, economic and environmental problems. It combined a critique of traditional notions of economic development with an equally trenchant critique of mainstream environmentalism – a critique which had already absorbed some of the most important lessons of the emergent environmental justice movement.

But in order to fully appreciate the merits of the Brundtland Report, as well as the peculiar pathos of its original title, one needs to contrast it not only with what became of sustainability in the course of its co-option, but also with the stark neo-Malthusian scenarios which dominated environmental discourse at the time – and which had primed my own teenaged self to recognise humanity in a box of maggots. The dire predictions of authors such as William and Paul Paddock (1967), Paul Ehrlich (1968) and Donella Meadows et al. (1972) challenged the belief that scientific progress and political emancipation went hand in hand. They suggested that efforts to eliminate poverty, starvation and disease were futile, if not self-destructive. The survival of the species necessitated that a large part of the world's human population be allowed to die (unsurprisingly, this was mostly meant to apply to those people who had historically been the primary targets for the exercise of necropolitical sovereignty; Mbembe 2003: 18–25). By the late 1970s, prominent thinkers such as Robert Heilbroner, Hans Jonas and William Ophuls openly entertained the possibility that an effective response to the crises of overpopulation and over-consumption might require that individual rights and democratic freedoms be substantially curtailed (Blühdorn 2011); famously, the Ehrlichs declared their support for China's coercive one-child policy (1990: 205). Only an authoritarian state, it seemed, would be able to impose the necessary sacrifices on its citizens.

Like the discourse of sustainability, neo-Malthusianism is not merely a set of descriptive statements or policy recommendations. It also touches on our conceptions of the human and its relationship to the larger world, on what we may hope and should fear, on the possible scope of our knowledge and our actions; it has an anthropological dimension, in the philosophical sense of that term. While my little larval epiphany may have reflected anxieties typical of a youth spent in suburban West Germany (in a household of lapsed Catholics, for that matter), I think that it also encapsulates, with the luridness of a dream image, some of the most unsettling anthropological implications of neo-Malthusian thought. It suggests that humanity's destructive impact on the natural environment is not at all a sign of our alienation from the natural order but, on the contrary, springs from our very naturalness; that the technologies our species has devised in order to dominate the natural world do not set it apart from the latter but, on the contrary, constitute merely an extension of basic biological imperatives. Instead of confirming my uniqueness as a person, my pursuit of individual desires marks me as just another wriggling body in the tupperware box. In the threat of overpopulation and over-consumption, human life encounters itself as reduced to sheer biology, as both de-individualised and de-socialised, and as utterly overwhelmed by its own inherited drive to reproduce and consume. Far from demonstrating that humans are somehow unnatural, our tendency to multiply beyond the Earth's carrying capacity is a sign that we are indeed a biological species just like any other, fully in the thrall of blind evolutionary forces that lie beyond our control. The story of human ascendancy is revealed as a kind of *Kippfigur*: a reversible image in which two contradictory meanings continually displace each other. The maggots are a figure both for the intolerability of purely biological life *and* for what humans are or have become in the very effort to extricate themselves from it. What, at one moment, looks like the emancipation of humanity from the vagaries of natural existence turns out to be, in the next, merely the passage to a higher and even more precarious level of unfreedom. The image inspires a paradoxical self-disgust – paradoxical because the self that judges both is and is not the self that is judged, both is and is not the maggot.

Thus understood, neo-Malthusianism represents something like an internal rupture in the master narrative of the Enlightenment – internal insofar as this rupture results not from a collision of its emancipatory programme with some extraneous obstacle, but rather as a consequence of the unfolding of its own logic. This logic is premised, as Foucault put it, on the linkage of 'the progress of truth and the history of liberty in a bond of direct relation' (1984: 43). From the seventeenth century onwards, the governing assumption had been that every increase in our knowledge of nature would yield a commensurate increase of human autonomy. The more we learn about the natural world, the greater the margin of freedom we enjoy in constructing a human world where the dignity of the individual can find respect. Neo-Malthusianism confounded this logic: in applying the principles of scientific naturalism to the human species itself, it had arrived at an account of the latter according to which scientific truth now demanded the curbing of emancipatory aspirations and a relinquishment of individual autonomy.

The Brundtland Report (and the discourse of sustainability which it inaugurated) can be seen as an attempt to close this rupture and to suture ecological science back to a politics of emancipation. Significantly, it opened with one of the most recognisable tropes of environmentalist discourse:

> In the middle of the 20th century, we saw our planet from space for the first time ... From space, we see a small and fragile ball dominated not by human activity and edifice but by a pattern of clouds, oceans, greenery, and soils. Humanity's inability to fit its activities into that pattern is changing planetary systems, fundamentally. (World Commission 1987: 1)

Here, the planetary ecology figures as a well-ordered structure that imposes fixed limitations which society must adjust itself to. Despite the image's distinctly pastoral overtones, this is the Earth of *The Population Bomb* and *The Limits to Growth*. However, already in the next paragraph, the writers insist that a respect for these limitations need not contradict the human drive for self-improvement: 'We have the power to reconcile human affairs with natural laws and to thrive in the process' (World Commission 1987: 1) The aim of sustainable development, the commission writes, is 'a new era of economic growth' which will '[extend] to all the opportunity to fulfil their aspirations for a better life' (World Commission 1987: 8). On the one hand, then, the report acknowledged the validity of neo-Malthusian concerns; on the other, it emphatically reasserted the idea of history as a gradual progression towards universal emancipation and prosperity.

The Brundtland Report has often been criticised for its failure to reject the goal of economic growth, which, it is said, stands in conflict with the idea of ecological sustainability. But in light of the foregoing, I want to argue that this is only one surface manifestation of deeper contradictions which the concept of sustainable development, in the effort to deflect the neo-Malthusian provocation, *contains* – in the double sense of comprising and obviating them. There is, first of all, very little warrant for the optimistic assumption that the goal of ecological sustainability makes a natural fit with emancipatory politics. On the contrary, there are ample grounds for believing that the push for greater individual autonomy and democratic participation, at least in the current understanding of these terms, works against the kinds of collective constraints and against the equitable sharing of environmental burdens across national boundaries which would be required in order to bring world society onto an ecologically sustainable path (Blühdorn 2011). The consistent tendency of liberal democracies over the past decades has been to expand the domain of individual autonomy. It is not only in the US that attempts by the state to constrain citizens in the freedom of their 'private' choices frequently meet with popular outrage (consider, for example, the German public's reaction to the Green Party's proposal, in the run-up to the 2013 elections, to introduce a compulsory 'veggie day' in cafeterias; Connolly 2013).

It is equally misguided to think of sustainability as a way of aligning society with natural principles. If, in the light of contemporary ecological science, it is still possible to speak of nature as a harmonious, stable, self-regenerating order, this characterisation can only apply at the highest levels. The ability of life as a whole to maintain dynamic equilibrium is predicated on perpetual flux, on the continuous destruction and rebirth of biological species and the individuals composing them. It must be remembered that Darwin himself had imagined Malthusian crises as the very dynamo of evolution. Although contemporary evolutionary theory has complicated this picture, there is much evidence to suggest that the selection pressure resulting from population overshoot played a key role not only in natural evolution, but also in human history (Christian 2014: 232–8). To the extent that the principle of sustainable development aims to forestall such crises in order to secure the prosperity and autonomy of individual humans, as well as the survival of the species as a whole, it aims not so much to fit human activities to the 'pattern' of planetary ecology, in the terminology of the Brundtland Report, as it attempts to eject humanity from this pattern altogether. After all, from the perspective of evolutionary biology, 'it is solipsistic nonsense to expect any fate other than extinction for *homo sapiens*' (Margulis and Guerrero 1989: 66). Insofar as sustainability subsumes economic, political and ecological rationalities under the single imperative of species survival, it can be said to 'naturalise' the human species; however, in doing so, it also evacuates nature of any normative content. If all forms of human behaviour are assessed, sanctioned and perhaps reconstituted with a view to the question whether they might impair 'the ability of future generations to meet their own needs' (World Commission 1987: 8), this effort could not produce a 'more natural' social order, as long as the word 'natural' retains any of its traditional connotations of spontaneous, self-directed growth. On the contrary, such a social order would be more thoroughly rationalised, and therefore denaturalised, than any other in the history of humankind; indeed, because everything that can be done in the present would already be circumscribed by a future anticipated in advance, such a society would have neither a history nor a politics in the customary sense of these words. A species which embarked on such a project would be radically different from any naturally evolved species, and the individuals belonging to it would no longer be humans as we know them.

To sum up the foregoing, one can say that sustainability contains a double paradox. It seeks to reconcile principles which are essentially at odds with each other – a neo-Malthusian conception of absolute natural limits with an emancipatory politics, a normatively charged conception of ecology as 'natural law' with the imperative of species survival. But this paradoxical quality should not be mistaken for a flaw. Sustainability articulates conflicting goals which are equally compelling, but neither of which could be realised completely without cancelling out the other. This is precisely the reason why the terms of sustainability are contestable: only because sustainability contains alternatives that are impossible to fully reconcile can it constitute a genuinely political matter (Mouffe 2000: 4–5).

Michel Houellebecq's modest proposal

The Possibility of an Island, originally published in 2005, is Michel Houellebecq's fourth novel, and it cemented his already established reputation as a literary provocateur. Much like his earlier novels, it features a fish-eyed vision of the psychological devastation wrought by consumer capitalism, graphic but disturbingly affectless sex scenes, dollops of philosophical and sociological speculation, and a deeply misanthropic protagonist resembling the author's own public persona. It also expands on and modifies the transhumanist theme Houellebecq had already introduced in *Atomized* (2000 [1998]). The plot of *The Possibility of an Island* alternates between the autobiographical account of Daniel1, which constitutes the bulk of the narrative, and the commentaries added to the latter, 2,000 years into a post-apocalyptic future, by his cloned descendants Daniels 24 and 25. Daniel1 is a French comedy star who has made a fortune with anti-human rights, anti-family, anti-Muslim, anti-Semitic, misogynist and generally offensive television shows (tellingly, one of his productions is called '100% hateful'; Houellebecq 2006: 45).[1] His story opens at the end of his thirties. Having grown tired even of his own disgust and with his sexual life in decline, he meets Isabelle, editor of the girls' magazine *Lolita*. They fall in love and marry. Isabelle is intelligent and ravishingly beautiful, but as she turns forty, she becomes increasingly self-conscious about her body, loses interest in sex, breaks off their relationship and eventually commits suicide. After their break-up, Daniel1 meets Esther, a Spanish student in her early twenties, who does not really love him but enjoys sex. Daniel1 falls hopelessly in love with her. As anticipated by Daniel1, Esther eventually loses interest in him, and he too commits suicide.

In the meantime, however, he has gravitated into the orbit of the Elohimites, a New Age cult closely modelled on the real-life Raëlians. The Elohimites promise their members unfettered sex and eternal youth through cloning. Due to his fame, Daniel1 is quickly introduced into the sect's inner circle. At a gathering on Lanzarote, shortly before his separation from Esther, he becomes witness to the events surrounding the heavily televised death and resurrection of the cult's prophet, a sham which marks the beginning of the Elohimites' triumphal ascendance during the following decades – an ascendance about which the reader learns, of course, not from Daniel1, but from Daniels 24 and 25, who belong to the race of neohumans. These are the eventual result of the Elohimites' eugenic experiments, most importantly the Standard Genetic Correction, which has equipped them, among other things, with the ability to photosynthesise. It is this change in their biological makeup, we are told (325), which has allowed neohumans to survive, without difficulties, the various cataclysms (involving nuclear wars, climate change, and a shift in the Earth's axis) that have all but wiped out the old human race, reducing it to scattered hordes of bestialised humanoids who have lost the capacity for language and are unable to maintain any complex form of social organisation. The neohumans no longer need to eat or defecate, they subsist on a diet of mineral salts, water and sunlight, and they no longer have sex. Their social contacts are reduced to infrequent videophone conferences. When a clone dies, his or her replacement – a mature individual with the body of an eighteen-year-old – is shipped from the Central City to the compound of the deceased within twenty-four hours. What these clones lack, however, is a complete memory of their predecessor's lives, as 'memory downloading through the intermediary of a data carrier' has turned out to be unfeasible (18); instead, they must read their 'life stories' and add their own commentary to them, as Daniel 24 and 25 are doing.

Neohuman society as it emerges from the comments of Daniels 24 and 25 is both the perfect consummation and a searing caricature of the ideal of ecological sustainability. By becoming autotrophic, neohumans have reduced their impact on the environment to an absolute minimum. By replacing sexual reproduction with cloning, they have completely stabilised their population at an exceptionally low number – there are never more than 6,174 neohumans at any one time (74) – and resolved the problem of intergenerational equity, to boot: we learn rather little about their political organisation, but from what we do glean, it is to be inferred that all neohumans live under very similar material conditions.

The just distribution of resources is, in other words, no longer an issue. Yet these achievements should not be misunderstood as 'political' in the conventional sense of the term – rather, they are for the most part consequences of the Standard Genetic Correction. Thus Daniel25 mocks Nietzsche's definition of man as '*das nicht festgestellte Tier*' ('the animal whose type is not fixed') which, he writes, was already false with respect to humans and is even less true for neohumans, whose society leaves as little room for individual variation as for social change. Since 'the root of all evil was biological' (124), human suffering cannot be abolished through political revolutions, but only by a transformation of human biology. Thus the Seven Founders, 'who created the Central City' and to whom the political organisation of 'neohuman communities owes almost everything', view politics as an 'inessential parameter' (370). In the '*Prolegomena to the Construction of the Central City*', they state as their goal the creation of an 'exhaustive cartography of life situations'; as Daniel25 tells us, the chief inspiration for this founding document of neohuman society was the manual for the 'video player JVC HR-DV3/MS' (392–3). Having accepted their status as biologically determined creatures, neohumans no longer have a history, properly speaking, nor do they have individual lives in our sense of the term:

> A limited calendar, punctuated by sufficient episodes of mini-grace (such as are offered by the sun slipping across the shutters, or the sudden retreat, under the influence of violent wind from the north, of a threatening cloud formation) organizes my existence, the precise duration of which is an indifferent parameter. Identical to Daniel24, I know that I will have, in Daniel26, an equivalent successor; the limited, respectable memories we keep of existences that have identical contours do not have any of the pregnancy that would be necessary for an individual fiction to take hold. The life of each man, in its broad brushstrokes, is similar. (371)

One might describe the transition from the catastrophic human to the sustainable neohuman era as a 'naturalisation' of history; at the same time, it obviously represents a radical break with human nature as shaped by biological evolution. The linchpin of neohuman sustainability is the abolition of sex through technical means; even if neohuman history seems 'peculiarly calm' (362), it is still oriented towards an eschatological horizon, namely the complete eradication of desire, which, according to the religious teachings of the Supreme Sister, will usher in the arrival of the 'Future Ones' (84), when 'the great sun of the moral law … would finally shine on the surface of the world' (422). The eradication of desire is also the purpose of the practice of reading and commenting on the predecessors' 'life stories': the goal is to cultivate 'repugnance and boredom' and to measure the distance that neohumans have put between themselves and the wayward desires which had ruled the lives of humans (84). Daniel1's 'life story' thus assumes the status of hagiography: Daniel1 is a martyr of biology, and with its complete rejection of human existence as lived in the contemporary world, his story foreshadows the coming moral order. With this in mind, the sexual antics of Daniel1 might perhaps best be compared to the orgies of the antinomian Gnostics, whose explicit purpose was to prepare the soul for return to its divine origin by humiliating the flesh (Lee 1987: 133).

While the life story of Daniel1 has indeed inspired disgust in many contemporary readers, it fails to have the proper effect on Daniel25 – rather than fortifying the indifference which the Supreme Sister has set up as the highest goal of neohuman life, it nourishes a 'nostalgia' for the desires that he is no longer able to feel (Houellebecq 2006: 371). He decides to leave his compound and sets out to search for a hypothetical community of humans or neohumans who might have discovered a 'new mode of relational organization' (377). Not surprisingly, given Houellebecq's programmatic pessimism, his quest fails. Daniel25 discovers that there is indeed nothing in the outside world to warrant his hopes: 'The world was there, with its forests, prairies, and its animals in all their innocence – digestive tubes on paws, with teeth at the end of them, whose life amounted to finding other digestive tubes in order to devour them' (406). The last pages of the novel find him moored on a shoal in the dried-out Atlantic, resigned to spending the remaining sixty or so years of his life in a vegetable-like stupor:

> I bathed for a long time under the sun and the starlight, and I felt nothing other than a slightly obscure and nutritive sensation. Happiness was not a possible horizon. The world had betrayed. My body belonged to me for only a brief lapse of time; I would never reach the goal I had been set. The future was empty; it was the mountain ... I was, I was no longer. Life was real. (423)

A gift one cannot reject

To be sure, neohuman society as depicted in *The Possibility of an Island* is not advanced as a remedy for the ecological depredations of contemporary consumer capitalism (only once in the novel does Daniel25 refer to 'ecologism', describing it as a 'strangely masochistic ideology' which appeared during the last centuries of human civilisation and, in its desire to protect nature, had 'greatly underestimated the living world's capacity for adaptation'; Houllebecq 2006: 395–6). Rather, it is conceived as a techno-religious solution to fundamental problems of the human condition – a solution which, moreover, ultimately fails to deliver on its promise. One can fairly speculate that among Houellebecq's reasons for reiterating a transhumanist scenario quite similar to the one he had developed previously in *Atomized* was that many readers had erroneously understood the latter as a positive utopia (see e.g. Varsava 2005). Few readers will repeat this mistake with regard to *The Possibility of an Island*: the present and the future of the novel are, each in their own way, equally repugnant. They are also tethered to each other in curious and instructive ways.

There is a striking discrepancy between the professed aims of the Elohimites and the neohuman society which is the result of their efforts. The Elohimites view themselves as a group of elect who have attained a higher plane of human evolution and therefore stand apart from mainstream society. However, Daniel1 leaves little doubt that the sect is a product both of and for consumer capitalism. One reason why he is so quickly accepted into the prophet's inner circle is his marketing acumen, and on several occasions he discusses how the sect's message should be tailored to reach its target audience more effectively. Elohimism, he suggests, is the avant-garde of that hedonistic tendency which has guided the entire 'movement of history ... since the end of the Middle Ages' (Houllebecq 2006: 366). After his first encounter with members of the sect, he begins to refer to them as 'the Very Healthy Ones' (97). The prophet adheres strictly to a 'Cretan Diet' of untainted, natural foods (201). About his followers, Daniel1 remarks that they:

> did not want to grow old; ... they forbade themselves from smoking, and took anti-radicals and other such things that you generally find in pharmacy shops ... Alcohol was permitted, in the form of red wine – limited to two glasses a day ... Health was the objective. All that was healthy, and therefore, in particular, all that was sexual, was permitted. (97–8)

The Elohimites' obsession with physical health reflects the priorities of a culture that has equated emancipation with the liberation of individual desire, and where the maintenance of the body's ability to experience pleasure has therefore become the paramount objective. The sect's promise of immortality appeals to 'the hope of an indefinite continuation of [an] existence that was devoted to pleasure' (366). Self-enhancement is conceived as a spiritual project. Their anti-natalist advertising campaigns, whose slogan is 'JUST SAY NO. USE CONDOMS,' lay the emphasis on what an unwelcome encumbrance children are for their parents (347). In their devotion to 'science, art, creation, beauty, [and] love' (217), the Elohimites believe themselves to be acting on natural principles and, at the same time, as advancing human freedom.

This refined form of hedonism makes for an almost perfect match with the LOHAS consumer profile. The ironic punchline of *The Possibility of an Island* is that the Elohimites' quest for sustainable pleasure finds its fulfilment in a society that, on the face of it, appears almost like a photographic negative of their vision. Instead of expanding the human capacity for pleasure, neohumans have ended up systematically eradicating the desire for it. In perpetuating individual life (there will always be a Daniel in a Mediterranean beach house), they have also evacuated its particular qualities and effectively rendered it indistinguishable from species survival. Instead of liberating the individual from social constraints, they have shattered society into a loose aggregation of monadic selves leading lives which render the very notion of freedom meaningless. And yet, neohuman society has not managed to slough off the problems it was designed to solve: human desire cannot, after all, be objectified, controlled and extinguished in the manner envisioned by the Seven Founders. Neohuman society ensures that Daniel25 is a stranger to physical want; still, he ends up:

> envying the destiny of Daniel1, his violent and contradictory journey, the amorous passions that had shaken him, whatever his suffering and tragic end. That immense joy, that transfiguration of his physical being by which Daniel1 was submerged at the moment of the fulfillment of his desires, … I had never known, I hadn't even any notion of them at all, it seemed to me now that, under these conditions, I could not go on living. (383–4)

It is the entanglement with the other, with all the pain it entails, which alone makes life worth living but must, in the end, also lead to its undoing. A life whose only purpose is to sustain itself is unbearable. If humans were to gain immortality and freedom from suffering, they would find them stale. If sustainability is to mean something other than the perpetuation of the same, it must fasten on the finitude of individual life as the necessary condition in order for that life to have value. Jos de Mul suggests that we recognise in Daniel25's envy the restatement of a very old theme: the jealousy of the Olympian gods for a happiness only mortals can know. He thus views *The Possibility of an Island* as articulating a renewed (and, he argues, uniquely European) 'tragic humanism' (de Mul 2014: 104).

Appealing as such a reading may be, I think that it underplays the corrosive, ill-tempered humour that is so characteristic of the novel. Houellebecq is clearly not interested in presenting a fair and balanced account of the human condition. However much Daniel1 may pride himself on his unconditional honesty, the picture he draws of contemporary life is a severely constricted, hopelessly tendentious one which methodically elides all its redeeming features, as John Updike has rightly pointed out (2006). The sustainable dystopia of his descendants does not represent a positive alternative to the unsustainable present. Given the novel's anthropological premises, it is only consequential that it would present its readers with such an impasse: the failure of the neohuman experiment, it suggests, reflects a flaw in human existence so fundamental that no amount of tinkering with human biology or social arrangements could possibly redress it. Jack Abecassis has argued that *Atomized* is informed by a *contemptus mundi*, a disdain for the world, inherited, by way of the French *moralistes*, from Pauline and Augustinian theology, but which Houellebecq has shorn of the promise of redemption through faith, such that only the extinction of humanity and its replacement by something radically different can offer hope (Abecassis 2000: 823–4). Much the same can be said for *The Possibility of an Island*, except that here even this last resort is walled up. The negation of the negative does not lead to affirmation, but only more deeply into despair.

So, surely, Houellebecq's crypto-Catholic nihilism, as such, cannot be of any help when it comes to the actual task of building more sustainable forms of life. However, it seems to me that this, by itself, does not set him apart from those who see sustainability as wedded to an ethos of existential affirmation. Indeed, one of the most striking aspects of the place in which Daniel25 finds himself at the novel's end is its discomforting similarity to the world of 'radical immanence' which Rosi Braidotti envisions as the horizon of posthuman becoming. He is a nomadic subject moving across a 'common life-space' which he 'never masters nor possesses but merely inhabits [and] crosses' – only that he is alone, and quite unregulated by an 'ethics of joy and affirmation' (Braidotti 2013: 193). This recalls a point Braidotti herself makes: that such an ethics rests on a 'fundamental gratuitousness' (2013: 192). If the imperative of sustainability entails the recognition that our entanglement with the world is absolute and without alternative, the call to affirm it is entirely supererogatory. At the same time, the question of whether this world is fundamentally hostile or hospitable to human flourishing assumes a desperate urgency. But this hardly means that an answer is readily at hand. That human life is fundamentally imbricated with the life process as a whole can be a curse or a blessing; the value of a gift one cannot reject is difficult to assess. *The Possibility of an Island* is a reminder of the radical ambivalence we are left with after the bond between truth and liberty which had defined the project of the Enlightenment is broken, and nature can no longer point the way towards the good life.

To search for a more sustainable way of life, then, is to negotiate provisional settlements between the conflicting claims of ecological science and emancipatory politics, of species survival and individual (as well as communal) autonomy, of definite limits and illimitable needs. Sustainability is the name we have given to this antagonism, rather than a first principle which could be invoked in order to resolve it. What is most admirable about a document such as the Brundtland Report is therefore neither that it successfully popularised the concept, nor that it hammered it into an intellectually appealing or politically robust shape, but rather the meticulous care with which its authors tried to lay out the dimensions of this field of struggle, to assemble as many of the concerned parties as they could, and to account for their competing visions of the good life. In the end, it seems to me that this effort might be the only crucial difference between their notion of the concept and the one articulated in Houellebecq's imaginary *Prolegomena to the Construction of the Central City*; but it makes for all the affirmation that sustainability needs.

Note

1 All references to *The Possibility of an Island* are to this edition, unless otherwise stated.

References

Abecassis, Jack I. 2000. 'The Eclipse of Desire: L'affaire Houellebecq', *MLN* 115 (4): 801–26.

Alaimo, Stacy 2012. 'Sustainable This, Sustainable That: New Materialisms, Posthumanism, and Unknown Futures', *PMLA* 127 (3): 558–64.

Blühdorn, Ingolfur 2011. 'The Sustainability of Democracy: On Limits to Growth, the Post–Democratic Turn and Reactionary Democrats', *Eurozine*. http://www.eurozine.com/the-sustainability-of-democracy/ Accessed 17 March 2017.

Braidotti, Rosi 2013. *The Posthuman*. Cambridge: Polity.

Christian, David 2014 [2004]. *Maps of Time: An Introduction to Big History*. Berkeley: University of California Press.

Connolly, Kate 2013. 'Wurst Policy Ever? German "Veggie Day" Plan Leaves Greens Trailing', *Guardian* 13 September. www.theguardian.com/world/german-elections-blog-2013/2013/sep/13/german-election-wurst-policy-veggie-day-greens. Accessed 21 February 2017.

De Mul, Jos 2014. 'The Possibility of an Island: Michel Houellebecq's Tragic Humanism', *Journal of Aesthetics and Phenomenology* 1 (1): 91–110.

Ehrlich, Paul R. 1968. *The Population Bomb*. New York: Ballantine.

Ehrlich, Paul R. and Anne Ehrlich 1990. *The Population Explosion*. New York: Simon and Schuster.

Emerich, Monica 2011. *The Gospel of Sustainability: Media, Market, and LOHAS*. Urbana: University of Illinois Press.

Foucault, Michel 1984. 'What Is Enlightenment?' *The Foucault Reader*. Ed. Paul Rabinow. New York: Pantheon, 32–50.

Houellebecq, Michel 2000 [1998]. *Atomized*. Trans. Frank Wynne. New York: Vintage.

Houellebecq, Michel 2006 [2005]. *The Possibility of an Island*. Trans. Gavin Bowd. London: Phoenix.

Keller, Lynn 2012. 'Beyond Imagining, Imagining Beyond', *PMLA* 127 (3): 579–85.

Lee, Philip J. 1987. *Against the Protestant Gnostics*. Oxford: Oxford University Press.

Margulis, Lynn and Ricardo Guerrero 1989. 'From Planetary Atmospheres to Microbial Communities'. In *Changing the Global Environment: Perspectives on Human Involvement*. Ed. Daniel Botkin et al. New York: Academic Press, 65–81.

Mbembé, Achille 2003. 'Necropolitics', *Public Culture* 15 (1): 11–40.

Meadows, Donella H. et al. 1972. *The Limits to Growth*. New York: Universe Book.

Mouffe, Chantal 2000. *The Democratic Paradox*. London: Verso.

Paddock, William and Paul Paddock 1967. *Famine 1975! America's Decision: Who Will Survive?* Boston: Little Brown.

Ray, Paul H. and Sherry Ruth Anderson 2000. *The Cultural Creatives: How 50 Million People Are Changing the World*. New York: Three Rivers.

Updike, John 2006. '90% Hateful: A New Novel by the Provocative Michel Houellebecq', *New Yorker*. 22 May.

Varsava, Jerry Andrew 2005. 'Utopian Yearnings, Dystopian Thoughts: Houellebecq's *The Elementary Particles* and the Problem of Scientific Communitarianism', *College Literature* 32 (4): 145–67.

World Commission on Environment and Development 1987. *Our Common Future* (Brundtland Report). Oxford: Oxford University Press.

11

Jorie Graham's *Sea Change*: the poetics of sustainability and the politics of what we're sustaining

Matthew Griffiths

In her 2008 collection *Sea Change*, US poet Jorie Graham pursues a concern about how language can engage with and represent material force, a concern that has preoccupied her in previous work. But *Sea Change* marks a distinct development of this in two key respects: not only does Graham adopt and sustain a particular form throughout the book to explore the tension between word and world, her concerns also inform a number of pieces that refer, albeit obliquely, to the concept of sustainability. The collection is described in the blurb as 'poetry of the tipping point, when what is lost and damaged in our world and our humanity is forever irrecoverable, when time itself has disintegrated' (Graham 2008a: back cover).

In the context of what is 'lost' and 'irrecoverable', sustainability is clearly a problematic concept, as can be seen in remarks Graham made reflecting on the book's composition: she writes that she 'realized that though it might indeed be, as many scientists think, "too late" to completely avoid an unsustainable world – such knowledge is both true, *and* baffling to the soul' (2012: 5; author's italics). A corresponding dilemma is summarised in the foreword to *Sustainability Education*, when Peter Blaze Corcoran writes that 'sustainability has become the metanarrative of our time – while at the same time sustainability has become a diminishing prospect' in actuality (2010: xiii). His analysis, like Graham's, identifies a discrepancy between the ideal of sustainability and our inability as a civilisation to achieve that ideal. This discrepancy arises because of the contingency of any definition of 'sustainability', a problem that is of increasing concern in the field. For instance, Leerom Medovoi, in his article, 'A Contribution to the Critique of Political Ecology: Sustainability as Disavowal', points to the 'substantial resources and interests behind the word' (2010: 129), while Kristiina A. Vogt et al. note in *Sustainability Unpacked* that 'defining sustainability has been challenging because of the need to include social, economic and environmental factors simultaneously' and that '[o]ne definition cannot and should not encompass the complexity and capture the nuances that are inherent in the word "sustainable"' (2010: 3). Nevertheless, as Medovoi indicates, '[i]t is taken for granted that "sustainability" refers specifically to the maintaining of something that is humanly valued' (2010: 130), pointing up the inherent anthropocentrism of the concept.

This tacitly self-interested definition also informs Medovoi's observation that sustainability 'has become a compulsively used word to get at some unspecified but ubiquitous notion of an environmentally ethical and conscious way of life' (2010: 129). For Vogt et al. the difficulty of specifying what is meant lies partly in trying 'to include social, ecological and economic factors with all their interconnections and possible feedbacks into one story. Many factors need to be included that may not, at first glance, appear to be relevant' (2010: 4). Medovoi's observation that the term 'sustainability' is 'compulsively used' and 'ubiquitous' corresponds with Corcoran's notion of sustainability as a 'metanarrative'; similarly, the difficulty that Vogt et al. perceive in telling sustainable stories represents the difficulty with which we make that ubiquitous notion a reality.

Given the tensions within 'sustainability' as an 'unspecified' term, and our need to establish 'interconnectedness' between 'many factors' which 'may not, at first glance appear to be relevant,' I will in this chapter propose that the poetics Graham employs in *Sea Change* makes use of the resources of poetic language, technique and tradition to engage with, define and take issue with conceptions of sustainability. In doing so, I contend that she challenges a previous model of 'sustainable poetry' advanced by Leonard M. Scigaj in his 1999 book of that title, depending as that does on a particular, relatively narrow definition of sustainability rather than taking up its problematic ambiguities. The inherent tensions of the term recur throughout *Sea Change*, and I will endeavour to illuminate critically the way in which the poet's technique addresses the politics of sustainability in the twenty-first century.

I will begin by examining Scigaj's concept of 'sustainable poetry' and suggest how it becomes increasingly problematic in the years after his book's publication; through a reading of the title poem of *Sea Change*, I will show how Graham engages with these problems while using techniques for which Scigaj criticises her earlier work. I will proceed to analyse the way Graham's engagement is sustained throughout *Sea Change*, and go on to examine how the book's dialogue with the literary tradition attempts to sustain our culture. I will then reflexively examine how Graham endeavours to sustain art in the face of twenty-first-century environmental change, highlighting the significance of the sustained sequence to our engagement with such change. Graham's poetics implies that sustainability, far from being an intentional political practice, is a fraught, contingent, but nevertheless persistent human habit.

The unsustainability of sustainable poetry

Graham's poetic practice challenges the ease with which we have adopted 'sustainability' as an 'unspecified but ubiquitous notion', in Medovoi's words. Scigaj himself adopts the term when he writes that '[s]ustainable poetry maintains a healthy balance between ... textual and referential needs', and he bases his understanding of sustainability on the 'principle of biocentric harmony', proposing that 'sustainability means that humans can harvest a sufficient amount of a natural resource for consumption so long as we do not deplete the resource base' (1999: 78). Already, this indicates the tacit anthropocentricity – 'humans can harvest a sufficient amount' – to which Medovoi draws our attention. Scigaj also invokes a specifically aesthetic notion of balance in the form of 'biocentric harmony', thus using cultural terms to define a presumed natural condition, on which he will in turn base his titular, cultural concept. When Vogt et al. use a similar, extended analogy, thinking 'of the global community as a "symphony orchestra"', they point out that it is only when we 'coordinate resource consumption globally' that it 'becomes more sustainable; that is, the "symphony" plays a beautiful piece of music' (2010: 9). So what for Scigaj is an originating principle is shown by Vogt et al. actually to require a complex effort of orchestration.

Scigaj proceeds from his definition to suggest that: '[t]he resource base of poetry is the referential world, and language that evades the referential world through divorcing text from context ... is not sustainable' (1999: 79). However, in proceeding from a material to a poetic sustainability, Scigaj examines the definition of neither. As Vogt et al. point out, just '[b]ecause we have defined sustainability, this does not automatically mean that we understand how to implement practices to achieve its goals' (2010: 4); there is, in contrast, no clear sense in Scigaj why sustainability as defined in material terms can only be implemented through referential poetics. For example, to consider a term such as poetic resources will suggest more than just a poet's referents, and will include the techniques and traditions at their disposal as well. So when Scigaj asserts that 'a poetic oeuvre driven by theory rather than the actual lived experience of the poet within the stubborn complexities of daily existence' is not sustainable (1999: 79), he then eschews what other, non-referential resources might offer in terms of an engagement with sustainability.

This much may be clarified by a pair of complementary critical observations. John Elder maintains in an interview in *The Clearing* that an 'authentic narrative is one that allows you to live sustainably in place, while an inauthentic narrative has the opposite effect' (2015); while Elder does not elaborate on these terms, he shares with Scigaj a concern that literature should help us live sustainably. To qualify this, we should also remember that the 'authentic' need not consist entirely in the referential.

Of Graham's earlier work, Helen Vendler writes in *The Given and the Made* that 'the word "true" for [her] does not mean representational accuracy or scientific accuracy alone; the true, for an artist, must involve the accurate transmutation of feeling into knowledge, perception into categorization' (1995: 103).

Although he writes nearly a decade before the publication of *Sea Change*, Scigaj also claims that Graham's earlier work demonstrates a 'poetics of textuality' that 'removes us from the practical world we must engage, moment to moment' (Scigaj 1999: 56). In *Sea Change*, however, Graham demonstrates that such a poetics can nevertheless be used to engage with the concerns that occupy Scigaj, and in so doing exhibits some of the qualities by which Scigaj characterises sustainable poetry. For example, Graham's work does 'emphasize the *relationality* of language – how it articulates and codifies ways that humans relate to nature and themselves' (Scigaj 1999: 32; author's italics), and – to an extent – 'does find ways to reassert and reinvigorate interconnections between the potential agent [i.e. the reader] and the referential world' (62). That Graham does so without abandoning the techniques for which Scigaj criticises her in *Sustainable Poetry* suggests that he may be too singular in his understanding of sustainability, but it also indicates how much more evident the complexity of the concept – and our failure to realise it – becomes in the nine years between the two books.

Sustaining sea change

That our metanarrative and experience are at odds is evident in the opening four lines of the title poem 'Sea Change' (Graham 2008a: 3). In these, the narrator first reports an unprecedentedly intense wind before comparing it with 'the recording' of weather and then remarking on its characterisation in 'the news'. The wind's very force is suggested by a blowing-back of sense at the first few line breaks, which mark transitions where we would not grammatically anticipate them – particularly in the enjambed 'Un- / natural'. This difficulty can only be countered by the sustained effort one requires to keep the meaning of the poem in mind as one reads. Yet the poem shows that human narrative still endeavours to contain and control the phenomenal world; we have a need for framing discourses that is evident in the explicit reference to 'the recording' of weather data and 'the news' that relays it. Notably, these occur before sensory confirmation, 'Also the body says it' (3). The priority of media over physical experience signals the ubiquity of discourse in our construction of the world, and in this capacity the poem reflects Graham's continuing concern with '[h]ow to give bodily perception its due in thought', which Vendler finds as 'already vexing' the poet's earlier verse (1995: 96).

What Scigaj refers to as 'actual lived experience', then, still requires a means of expression 'to evoke what one cannot completely convey in language, but can experience fully in the lived moment' (Scigaj 1999: 68). Graham's mediating discourse is far removed from the context of 'wilderness experience' or 'wild being' that Scigaj is keen for ecopoets to evoke (1999: 68), yet 'Sea Change' could still be said to report on the 'actual lived experience' of many who may not have the opportunity or inclination to 'gain a sense of wild being' in this way. Again, Graham's poetics become apposite to describe the difficulties of sustainability because of qualities she sustains from her earlier work, which, as Vendler remarks, show that we 'cannot assert the sort of mastery over experience' that would allow 'choosing to stay the fair moment for inspection' (1995: 128).

For Scigaj, the 'intrusive media' in Graham's poetry signifies that 'the only relief from corrosive materialism is lonely anthropocentric introspection' (1999: 59). However, by consistently enforcing breaks that do not coincide with syntactic pauses, Graham's versification in 'Sea Change' draws attention to its own artificiality, highlighting that our environment comes to us predominantly mediated and our understanding of it arrives at 'the body' only after it has been processed through weather records and news. As such, her technique is far from solipsistic, as Scigaj suggests, but indicates her recognition of the way experience is entangled with its mediation; Vendler notes that the poet's 'form mirror[s] the unstoppable avalanche of sensations and the equal avalanche of units of verbal consciousness responding to those sensations' (1995: 106).

This formal 'avalanche' suggests that we are in fact struggling with the sustained momentum of experience and in turn failing to sustain our framing discourses, our metanarratives. Graham's own use of the word '*sustained*' in 'Sea Change' is defined as 'in a hatred of / a thought' (2008a: 3; author's italics). What then seems to be sustained in this context is an intellectual resistance to the knowledge that the earlier 'news' conveys, a hatred of thinking it. The subsequent phrase, 'or a vanity that comes upon one out of / nowhere', is syntactically ambiguous, so it remains unclear whether the vanity is itself another '*sustained*' quality or is actually another object of the preceding 'hatred'. Simultaneously, then, this spontaneous 'vanity' is akin to the resistance to thought, and in being hated, also an anti-intellectualism that prompts self-disgust. While this may suggest 'the tortuous recesses of introspection' that Scigaj criticises in Graham's earlier work (Scigaj 1999: 59), the ambivalence is a necessary recognition of the 'vanity' that obtains in the concept of sustainability, that is, in trying to sustain 'actual lived experience' in the face of a thought that might threaten us.

In this context, we are not actively sustaining our culture; rather, we are passively sustaining an onslaught, in the manner that Medovoi compares to sustaining an injury: '[t]o "sustain" something can also mean to endure or withstand it', and so '[i]t suggests damage that we are not so much trying to eliminate as to find a way to survive' (2010: 131). Medovoi describes this as 'a striking definition precisely because it inverts the valence' (2010: 131) of the other, 'ubiquitous' definitions of sustainability to which we are accustomed. Such an understanding of sustainability reveals it as a position we are forced to adopt, rather than one that allows us a capacity for decision. Far from being sustained through time, this conception of sustainability only begins with the recognition that we must consciously begin defending our lives and culture against encroaching environmental crisis.

Indeed, in 'Sustainable This, Sustainable That', the critic Stacy Alaimo makes the following comments on Graham's lines: '[t]he abrupt departure of a sense of permanence may provoke the desire to arrest change, to shore up solidity, to make things, systems, standards of living "sustainable"' (Alaimo 2012: 558). Scigaj's diagnosis that 'Graham has an inadequate sense of agency' and that she 'cannot affirm our human potential for positive social action in the referential world' (1999: 59) would seem to be borne out in these later poems, because she identifies our reaction as both belated and self-centred. However, her poetic practice is then able to give expression to the tendency in sustainability discourse that identifies the gulf between what we think we can achieve and what we have achieved, which has only widened in the years since Scigaj's book was published.

A sustained experiment

Graham's distinctive style of versification throughout *Sea Change* enacts the contradictions inherent in considerations of sustainability by the tension she creates between the sense and the sound of the poetry, in particular its momentum. This style takes the shape of poems beginning with a line ranged left, sometimes extending across the width of the page, but on occasion finishing before halfway. That line is followed, in most instances, by between one and nine shorter, indented lines, which keep a consistent left-hand margin about 40 per cent of the way across the page. These are followed by another long line ranged left, then more, shorter lines, maintaining the secondary margin at roughly two fifths of the page width. There are no stanza breaks, while the syntax tends to be continuous and most lines are enjambed.

Visually, this creates a concentration of shorter lines some way in from the left-hand margin, and together with the lead-in of the fuller lines and the absence of stanza breaks, this creates a largely vertical momentum as the eye is drawn down the page. The use of punctuation, particularly parenthetical dashes, is syncopated with this, to create a tension between visual momentum and semantic hesitancy.[1] The continued movement of the reading eye across the page that the longer lines require, or the sustaining of breath when read aloud, make the poems provocatively rather than evocatively sensory; that is, the spaces make apparent the effort we have to sustain in reading, in a way that continuous prose, or even conventional verse forms, would not.

Graham's technique enacts the tension between an overall design and the difficulty of sustaining the pace that this design requires – replicating the discrepancy between the metanarrative of sustainability and our experience of failing to realise it – and also enacts the tension between uncontainable material phenomena and the human attempt to manage them. The failure to keep pace with change is evidenced in the remark, in 'Sea Change': 'how the future / takes shape / too quickly' (Graham 2008a: 3). The enjambed lines once more create breaks where we do not syntactically expect them, encouraging us to take breaths at the same time as forcing us to read through them to provoke the sense of a future 'taking shape' too quickly for us to control.

The processes of nature cannot be contained by form or syntax, however, and conventional categories are exceeded by the enjambed lines. As a result, when the narrator insists that a 'calm and / true' state 'did exist just yesterday' (Graham 2008a: 3), it reads as another projection of human order rather than as an affirmation of former certainties, further undermined by the improbable precision of 'just yesterday', rather than just 'yesterday'. The suggestion is that we are sustaining, or trying to sustain, the state of yesterday, circling round on our nostalgia to recreate our imagination of a past ideal. As in Alaimo's analysis, sustainability confers a belated value on practices of the past, a recognition that we only try to preserve these once it is too late to do so effectively.

While Scigaj contrasts Graham's earlier work with his notion of 'sustainable poetry' on the grounds that the poet 'finds comfort only in intrapersonal reverie' (Scigaj 1999: 58), it is only through the critique prompted by such reflection as her poetics enables that we begin to identify the relative novelty of this conception of sustainability. Graham thus sustains techniques for which Scigaj criticised her, to show that these can nevertheless offer distinctive insights into the same concerns that occupied him. She recognises that we are attempting to sustain a sense of ourselves and our cultural practices rather than our environments, which are in turn sustaining change at their own momentum even as they also sustain the impacts of human activity.[2]

Graham engages with human attempts to sustain an anthropocentrically defined world throughout the collection. The 'vanity' to which she refers in 'Sea Change', for instance, is again addressed in 'Belief System' (Graham 2008a: 45–7), a poem that explores the thoughts and ways of thinking that our present moment should supersede. In the piece, an anthropocentric exceptionalism is evident from the fourth line onwards: 'By *the mind* we meant / the human mind' (Graham 2008a: 45; author's italics). Graham's adoption of the collective voice, 'we', suggests how complicit we all are in this process: it becomes something with which we can identify. By sustaining this cultural introspection in 'Belief System', Graham sees through it to what it in turn sustains: 'Thinking was the habitation of a / trembling colony, a fairy tale—of waiting, love—of / the capacity for / postponement—'. First, here, is the intimation of how precarious our civilisation is – 'a / trembling colony' that we inhabit. But this is followed by the 'fairy tale' the 'colony' tells itself, and while 'fairy tale' may seem dismissive of our self-imposed narratives, Graham goes on to elaborate that we also inhabit such sympathetic qualities as 'waiting, love', and these in turn sustain 'the capacity for / postponement'. This 'capacity' is itself postponed first by the line break and, after it, by an indent, these lines being the first two of five with an indented left-hand margin of the kind described above, so the versification again enacts the human contrivance of putting things off. It is a thought that postpones, and a postponement of thought.

When thought is forced to confront '[t]he future', it asks: 'How could it be performed by the mind became the / question—how, this sensation called tomorrow and / tomorrow?' (Graham 2008a: 45). The attempt to perpetuate human forms of understanding and negotiation of the world – as in this instance performance – again acknowledges what Alaimo calls '[t]he human-centered discourses of sustainability' (2012: 562). Scigaj maintains that certain poets 'privilege private self-exploration' and 'merely allow the wasteful imperialism of dominant cultures to continue' (1999: 56); yet Graham suggests that sustainability itself represents an attempt to perpetuate 'dominant cultures', and without such broader, cultural 'self-exploration', we would be unable to identify as much.

Sustaining the canon in *Sea Change*

The 'question' Graham asks in 'Belief System' is one that concerns making the nebulous concept of the future meaningful at the experiential as well as intellectual level. But the incommensurability of the future with human experience is attested by Tom Cohen, who observes that climate change – to consider one aspect of twenty-first-century environmental change – signifies 'incompatible referentials arriving that would operate beyond archival memory and social history' (2012: 24). That is, we cannot simply seek to sustain former ideas in the light of environmental change. Nevertheless, Graham's allusion to *Macbeth* (5.5.19; Shakespeare 1984: 153) in 'Belief System', 'tomorrow and / tomorrow', suggests that our engagement with the future is enabled by an engagement with, or 'performance' of, the past; though, as per Cohen's observation, the past offers no true precedent for its 'tomorrow', and our notion of sea change itself experiences a sea change.

Quoted from Ariel's song in *The Tempest* (1.2.396–403; Shakespeare 2011: 200), the phrase 'sea change' connotes an irreversible alteration of a state of affairs, a paradigm shift; read literally, it even denotes a process of oceanic restlessness and mutability, such as Graham has grappled with throughout the book. At the same time, as a quotation, 'sea change' is an expression sustained in our language for four centuries by a tradition continued in the act of quotation itself. Graham's use of the phrase for her title thus enacts the recognition of unprecedented change while trying to sustain the culture that is jeopardised by it. As a poem, and throughout the book to which it gives its name, 'Sea Change' sustains a number of references to the canon that help organise and inform Graham's response to the changing global environment of the twenty-first century.

In particular, she recognises the way that the tradition itself changes through time; any act of sustaining is also necessarily a contextually dependent change. Sustaining the past is thus a recycling of it for a different function or purpose. This sustaining of tradition enacts T. S. Eliot's notion, as formulated in his 1919 essay 'Tradition and the Individual Talent', that the appearance of a new work of art will alter all those that came before it, adjusting the relationship that already existed between them (Eliot 1975 [1919]: 38). His implication is that innovative work makes us perceive the tradition in a fresh and distinctive way.

Eliot conceives of this relationship in orderly, sculptural terms, whereas it is productively compared by John Elder to Gary Snyder's notion of culture as being akin to 'recycling dead biomass' (Snyder quoted in Elder 2000: 228) – an altogether more organic analogy, apt insofar as it suggests how poetic material is broken down and reconstituted as new work. Nevertheless, although Elder sees culture as a process, 'something which one does' or 'a dynamic continuity' (2000: 229–30), Snyder's organic analogy obscures the intentional, creative act of engagement by a contemporary poet in sustaining poetry of the past, which Eliot's aesthetic analogy allows. Whether we choose Eliot's model or Snyder's, it is still clear that we establish a relationship with something more remote in time than Scigaj's 'actual lived experience', as Snyder comments that culture works with material 'derived not from grazing off the annual production of biomass, but from recycling dead biomass, the stuff of the forest floor, the trees that have fallen, the bodies of dead animals' (quoted in Elder 2000: 228).

The process of adopting and adapting past cultural practices is anthropocentric inasmuch as it looks to human precedent to deal with contemporary predicaments. In doing so, though, it seeks to sustain a tradition older than our present practices, or even the 'calm and / true' state '[w] hich did exist just yesterday' (Graham 2008a: 3). Vogt et al. correspondingly look to past cultures for sustainable precedents, and they write of agricultural communities in which '[k]nowledge about the edible foods that they gathered' was relayed from one generation to the next, 'building a storehouse of information over the centuries' (2010: 31). Eventually, then, '[h]uman history is a series of stories documented, in part, by people who were successful in adapting to their environment' (Vogt et al. 2010: 283). This conception of sustainability is not simply the panicked attempt to prolong the present, such as that described by Alaimo, but something that has endured as culture endures. It looks to the past but, in order to deal with the 'incompatible referentials' to which Cohen refers, adapts a long-term view to current requirements.

In accordance with this principle, Graham's adoption of Eliot sees a further adaptation of tradition because it suggests we can no longer form an order through the creation of new works of art, even retrospectively. As Alaimo alludes in identifying our desire 'to shore up solidity' (2012: 558), our vision of cultural continuity becomes the seeming chaos of Eliot's poem *The Waste Land* (2015 [1922]: 53–77) because our changing planet exposes contemporary culture's lack of internal cohesion.[3] Apart from the commonality of reference to Shakespeare,[4] what *Sea Change* also shares with *The Waste Land* is a tension between the human attempt to maintain order and vital, persistent material forces. When 'the future / takes shape / too quickly' in 'Sea Change', it is figured as 'grasses shoot[ing] up, life disturbing life' (Graham 2008a: 3); these echo the blooms emerging at the start of part I of Eliot's poem, 'The Burial of the Dead' (2015 [1922]: 55). Eliot manages to half-contain natural energies with the present participles that end the first three lines of his poem, suggesting a circular pattern even with the onward thrust of those parts of the verb. He creates a cycle from processes that go beyond the containment of the line, managing to keep growth temporarily in check. By the twenty-first century, even this momentarily sustained equilibrium is impossible, and Graham's form instead signifies the self-sustaining, runaway character of natural processes.

The force of the wind images this quality in both poems, and the comparison highlights the exacerbation of climate in the eighty-six years between them. In 'A Game of Chess', the second part of *The Waste Land*, the wind remains beyond a door, figuring the disturbance of the narrator's interlocutor (2015 [1922]: 59). As order increasingly disintegrates throughout *The Waste Land*, however, its final section is exposed to the elements, so we hear 'What the Thunder Said'. The wind in Graham's poem resembles Eliot's thunder in that it cannot be shut out, but it is given voice from the very beginning of her book. In 'Sea Change', the wind's voice refutes the claim that we are unaware of our participation in worldly phenomena: 'consider your affliction says the / wind, do not plead ignorance' (Graham 2008a: 3–4). As in *The Waste Land*, civilisation wishes to sustain itself in separation from environmental change, hence the distress expressed at the opening of Eliot's poem on the return of spring after winter (2015: 55).

But we can only maintain the state that Graham calls 'ignorance' by suppressing the continuity between past and present. This condition is then forced to confront its own artificiality in Graham's poem – 'away leaks the / past, much farther than it used to go, beating against the shutters' (2008a: 4) – while in *The Waste Land*, the attempt to bury the past beneath ground and ice is met with the recurring reassertion of its presence, whether as the flowers that open 'The Burial of the Dead', or the dead themselves that end the section (2015: 55, 57). In Graham's poem, the cumulative past of human interaction with the environment is imagined not as the dead but as the weather, 'beating against the shutters', but our resistance to it is still marked by the failed enclosure of human domestic space apart from nature.

The wind's imperative 'consider' is repeated later in 'Sea Change': 'Consider / the body of the ocean which rises every / instant into / me' (Graham 2008a: 4). This recalls the verb that directs our regard to the drowned sailor of 'Death by Water', part IV of *The Waste Land* (2015: 67), but a tonal shift between the two poems is seen in the way Graham writes as a first person, 'me', who is subject to the elements, rather than Eliot's symbolic Phlebas. That is to say, we cannot project human experience of the sea into a separate, impersonal figure, but we must deal with it in the first person. Furthermore, there is only a versified – that is to say, artificial – boundary in Graham's poem between 'ocean which rises every instant into' and 'me', so the environment impinges on personal experience. We are ourselves sustained by water, in the sense that Medovoi defines as 'to "furnish with the necessaries of life"' (2010: 130), but water's own significance exceeds this function.

Graham recognises the difficulty of sustaining a boundary between the self and its material surroundings. Her poetry was already characterised by such trespass of the environmental on the territory of the personal. Of Graham's earlier work, Vendler writes: '[t]he self must now portray itself *in* primary matter' and '[y]et the indifference of the material universe to our fate makes us hesitate to appropriate the phenomena as adequate symbols of ourselves' (1995: 125; author's italics). By sustaining this concern until and through *Sea Change*, Graham's poetics demonstrates the fallacy of maintaining a discrete sense of self as though it is separate from natural forces.

Sea Change's allusions to *The Tempest* are further developed in the poem 'Full Fathom' (2008a: 30–1), which begins with the sea but churns in the experiences of everyday living, so as to evidence not only a comprehensive breakdown in categories but also a collapse of scales and contexts. Graham is then able to bring wider environmental and cultural phenomena into personal proximity in the poem with the reversal of

Shakespeare's formulation: 'Those are pearls that were his eyes' (*The Tempest* 1.2.399; Shakespeare 2011: 200) becomes 'those were' / 'that are' in 'Full Fathom' (Graham 2008a: 31). The phenomena are rooted in an experiencing subject, 'his eyes,' to become a vision of human entanglement with 'carbon sinks' and 'carbon sources' (Graham 2008a: 31). Human institutions such as 'reparation / agreements, summary / judgments' are not then sustained as such; rather, their dependence and impacts on the environment are exposed through their position in this sequence of phrases. The poem also questions the attempt to sustain the present:

> when was it
> in your admittedly short
> life you
> were permitted to believe that this lasted
> forever[?] (Graham 2008a: 30)

Throughout, the poem's deferral of syntactic closure represents both the belief that we can sustain present conditions forever and its simultaneous fallacy. The colons and ampersands that punctuate the opening lines recur throughout the poem, which is also strung through with em-dashes to put off a full-stop until the end of the final line. The syntax itself is trying to sustain things here, as a string of recapitulations, from which narratives emerge but are only hesitant and divergent.

To stress the physical implication of human beings in the climate, Graham also redeploys the imagery of *The Waste Land* in 'Positive Feedback Loop' (2008a: 42–4), itself a term naming a self-sustaining phenomenal cycle. The poem moves from a meditative, attentive opening to imagine the titular process, a way of describing a change that, once instigated, sustains itself. Using one of Eliot's key symbols, Graham freights contemporary personal experience with the environmental processes that are beyond our grasp, both physically and mentally, when she invites us to use his image of dust (Eliot 2015 [1922]: 55) as a tactile model for ocean circulation, making the original spiritual connotations of that dust materially manifest. This represents a further engagement with the concerns Vendler identified in Graham's earlier work, of '[h]ow to give bodily perception its due in thought' (1995: 96), except that, here, Graham is trying to give thought its due in bodily perception in order to engage abstract phenomena through sensory experience. (We might compare Graham's poetic achievement of this with the fictionally established relation between the phenomenological and the speculative real in Yann Martel's *Life of Pi* as discussed by Louise Squire later in this book.) Graham's technique affords a similar value to experience as Scigaj does, but does so because of the broader world to which imagined experience can give us access, rather than experience for its own sake.

The lines of the poem again run across the page in a manner that demonstrates the difficulty of being able to follow the instructions 'try /

to hold in mind the North Atlantic Deep Water' and 'try to hold a / complete collapse, in the North Atlantic Drift' (Graham 2008a: 42), as we might try to retain the dust we are handed. The conceptually difficult – 'try / to hold in mind' – becomes what is physically impossible – 'try to hold a / complete collapse'. Graham's poetics successfully establishes a connection between failure at the experiential level and the failure of climatic mechanisms that play their part in sustaining human existence. Furthermore, the belatedness of what it is now possible to sustain is hinted at in the substitution of dry substances – first 'sand', then 'dust' (Graham 2008a: 42) – for the vast bodies of water we are asked to imagine.

A sustained note

While *Sea Change* reveals quite how problematic it is to sustain thought or a contemporary sense of culture, the poetry suggests an alternative, provisional possibility. The terminology of music runs through the collection – for instance, in the 'chorusing in us of elements' (Graham 2008a: 4), 'Who is one when one calls oneself / one? An orchestra dies down' (42) and 'The score does not acknowledge / the turner of / pages' (45). These images suggest that identity is sustained as a function of many participating agents in a concerted effort. Like the orchestra, humanity is able to create a harmonious world, and Graham's extension of the musical image across separate poems itself enacts that context of cumulative creation through multiple recurrences. Moreover, Graham's use of music as a motif – a non-mimetic art form – affirms the value of poetic technique to express the tensions of sustainability, using sensation as much as representation or argumentation.

It does take the effort of a collective, however, to create and sustain this fictive harmony: as Vogt et al. point out, sustainability at the global level would itself require such a co-operative, orchestrated effort by humanity (2010: 9). Because individual conceptions of ourselves are implicated in environmental change – as 'Full Fathom' indicates with its sequencing from the personal through the political into the phenomenal (Graham 2008a: 31) – when we revert to such individualistic conceptions, the orchestral effect becomes impossible. The assertion of individualism in 'calling oneself one' in 'Positive Feedback Loop' means in contrast that the music 'dies down' (Graham 2008a: 42). Another sense of 'sustained' is invoked here, albeit in its absence, that of a musical note being held.

This motif of music is central to 'The Violinist at the Window, 1918' (Graham 2008a: 32–4), a quasi-ekphrastic poem named for a painting by Matisse that involves visual art as much as literature and music in the creation and sustaining of culture. Speculating on the music that the violinist will play, the poem seeks 'the sustained one note of obligatory / hope' (Graham 2008a: 32). This 'obligatory hope' is further qualified by Graham's subsequent description of it 'taken in, like a virus', eventually processed by the body to be made 'natural again' (2008a: 32). As individuals, she suggests, we first resist what art conveys to us – perhaps in the way 'a hatred of / a thought' is '*sustained*' in 'Sea Change' (2008a: 3) – before being infected by it and coming to regard it as natural.

Danger arises, though, when we do not see that we have created the world and instead naturalise our conception of it. This would be to sustain a conventional idea of ourselves without acknowledging our active role in doing so. When 'the mind is hatched and scored by clouds' (Graham 2008a: 33), the verb 'scored' plots the meteorological phenomena like musical notes, giving them artistic form. But having naturalised an aesthetic view of the world, we assume an unwarranted power to preserve it indefinitely, and the narrator declares 'what is weather—when it's / all gone we'll / buy more', ensuring 'ages that shall not end' (Graham 2008a: 33). That envisages a perpetual recycling of the present rather than the 'dynamic continuity' that Elder suggests constitutes culture (2000: 230). When Graham then returns to 'the note, sustained, fixed' at the end of the poem, she hears it as a 'high note trembling—it is a / good sound, it is an / ugly sound' (2008a: 34), reflecting its essential ambivalence, our gradual acclimatisation to the notion conveyed in the note, but one still underscored by dissonance. All this occurs in a poem where 'the war to end all wars has come / to an end', but is then immediately undercut by the wry aside '—for a while' (Graham 2008a: 32). The poem takes place in an interlude framed by the world wars as the violinist is framed by the window and doubly by the frame of the picture, signifying our artificial, aestheticised containment of the present moment.

Given the cumulative nature of *Sea Change* as a book, this concept of a precarious interlude – the 'note trembling' of 'The Violinist at the Window, 1918' resonates with the 'trembling colony' that we inhabit in 'Belief System' – is taken up again in the final poem, 'No Long Way Round' (Graham 2008a: 54–6). This offers a coda to the motifs and themes of the collection, again contrasting the incommensurability of global environmental crisis with everyday experience. In two passages of this last poem, the verse clumps into a pair of paragraphs resembling prose. The first of these, beginning 'It is an emergency actually', centres on the break 'the whole 15,000 years of the inter- / glacial period' (Graham 2008a: 55), accentuated by the shortfall of the first of these lines compared to those preceding it, and by the lengthy indentation of the subsequent one. If we mark the interlude here with silence, it reminds us of the brevity of our geological window, the current 'inter- / glacial'; if we instead mark it by holding the reading breath, we realise the physical difficulty of sustaining even one unspoken line. In either event, the effect reminds us of our physical implication in the world. Graham's resumption with a further prose-like stanza creates an illusion that things are close to normal, this 'waking and doing', 'the getting done' (2008a: 55), but the interruption serves to communicate the contingency of our quotidian lives. The poetry creates a prosaic effect for the everyday, only to disrupt it with a poetic break that highlights the difficulty of sustaining an accustomed normality.

'No Long Way Round' draws in with a moment of seeming lyrical meditation that again makes this quality of a sustained narrative apparent, the 'need to tell / your story' (Graham 2008a: 55–6). Graham shows how that 'story' is confined to its profoundly human significance; it begins personally – 'how you met, the coat one wore' – and even when it approaches a more global scale, as in 'Positive Feedback Loop', it tries to contain this, limiting it to 'that part of / the planet' and 'the first Spring after your war' (2008a: 56).[5] Finally comes the desperation to sustain an imagined normality and hold on to it through the imagined restatement of the word: 'thousands of times / you want to say this—normal—'. In contrast to these staccato, insistent remarks, the final lines of Graham's poem remind us: 'there are sounds the planet will always make', that is, sustained notes, 'even / if there is no one to hear them' (2008a: 56). These resonate with the sustained, 'ugly' note of 'The Violinist at the Window, 1918' (Graham 2008a: 34) to make a sound that persists even though it can only be prospectively conceptualised, and never humanly experienced.

Beyond sustainability

Graham's technique across *Sea Change* does not so much take argumentative issue with the concept of sustainability – a concept exemplified in Scigaj's approach – as demonstrate the inherent difficulty of satisfying its contending definitions, discussed by Vogt et al., which are simultaneously ambitious and impractical. Instead, Graham's verse reveals that we tend to perpetuate an idealistic metanarrative such as the one Corcoran identifies while we endure environmental change, where both these verbs, 'perpetuate' and 'endure', represent distinct definitions of sustainability by Medovoi.

But Graham's poetics also shows that, in seeking to survive through what is to come, we must be alive to culture beyond the individual, beyond the present, and recognise – and make best use of – tradition, intention and contingency to comprehend our position. Produced in the tension between her poetry's momentum and the transitions she describes between the global and the banal are themes and motifs that operate according to their own aesthetic syntax, accumulating into a willed change that also recognises the weight and agency of what it changes to bring the work together across history and geography. This is the distinctive achievement of *Sea Change*, a design that sustains cultural tradition and collective endeavour even as these sustain the onslaught of environmental change in the twenty-first century.

Notes

1 This latter quality in particular is reflected in Graham's own measured, almost hesitant performance of the poems: see, for example, the clip 'Jorie Graham and Yusef Komunyakaa at the 92nd Street Y', in which she reads 'The Violinist at the Window, 1918' from *Sea Change* (Graham 2008b).
2 The impacts of our cultural practices will become more evident in the poem 'Full Fathom' (Graham 2008a: 30–1), discussed below.
3 Alaimo's phrasing echoes line 430 of the poem (Eliot 2015 [1922]: 71).
4 In a *Guardian* review of *Sea Change*, M. Wynn Thomas remarks '[s]ignificantly, this volume's title points us not to the redemptive vision of *The Tempest* but to [it] as ominously refracted through Eliot's *The Waste Land*' (2008).
5 This also echoes the underground vernal stirrings in *The Waste Land*, written in the shadow of its own war, as well as the interwar interlude of 'The Violinist at the Window, 1918'.

References

Alaimo, Stacy 2012. 'Sustainable This, Sustainable That: New Materialisms, Posthumanism and Unknown Futures', *PMLA* 127 (3): 558–64.

Cohen, Tom 2012. 'Introduction: Murmurations – "Climate Change" and the Defacement of Theory'. In *Telemorphosis: Theory in the Era of Climate Change*, Vol. 1. Ed. Tom Cohen and Henry Sussman. Ann Arbor, MI: Open Humanities Press, 14–42.

Corcoran, Peter Blaze 2010. 'Foreword: Sustainability Education in Higher Education: Perspectives and Practices across the Curriculum'. In *Sustainability Education: Perspectives and Practices across Higher Education*. Eds Paula Jones, David Selby and Stephen Sterling. London: Earthscan, xiii–xv.

Elder, John 2000. 'Culture as Decay: Arnold, Eliot, Snyder'. In *The Green Studies Reader: From Romanticism to Ecocriticism*. Ed. Laurence Coupe. London: Routledge, 227–34. Reprinted excerpt from Elder, John 1985. *Imagining the Earth: Poetry and the Vision of Nature*. Chicago: University of Illinois Press.

Elder, John 2015. 'An Interview with John Elder' by Isabel Galleymore, *The Clearing* 13 March 2015. www.littletoller.co.uk/the-clearing/an-interview-with-john-elder/. Accessed 19 March 2017.

Eliot, T. S. 1975 [1919]. 'Tradition and the Individual Talent'. In *Selected Prose of T. S. Eliot*. Ed. Frank Kermode. London: Faber, 37–44.

Eliot, T. S. 2015 [1922]. *The Poems of T. S. Eliot*. Vol. 1: *Collected & Uncollected Poems*. Ed. Christopher Ricks and Jim McCue. London: Faber.

Graham, Jorie 2008a. *Sea Change: Poems*. Manchester: Carcanet.

Graham, Jorie 2008b. 'Jorie Graham and Yusef Komunyakaa at the 92nd Street Y'. 92nd Street Y, New York. 16 October 2008', www.youtube.com/watch?v=i6JvxtseAi0. Accessed 21 February 2017.

Graham, Jorie 2012. 'Jorie Graham', *Poetry Book Society Bulletin* (Autumn): 5.

Medovoi, Leerom 2010. 'A Contribution to the Critique of Political Ecology: Sustainability as Disavowal', *New Formations* 69: 129–43.

Scigaj, Leonard M. 1999. *Sustainable Poetry: Four American Ecopoets*. Lexington: University Press of Kentucky.

Shakespeare, William 1984. *Macbeth*. Ed. Kenneth Muir. London: Methuen; Arden Shakespeare, 2nd series.

Shakespeare, William 2011. *The Tempest*. Ed. Virginia Mason Vaughan and Alden T. Vaughan. London: Routledge, rev. edn; Arden Shakespeare, 3rd series.

Thomas, M. Wynn 2008. 'Full Fathom Five'. Review of *Sea Change* by Jorie Graham. *Guardian* 3 May 2008. www.theguardian.com/books/2008/may/03/featuresreviews.guardianreview25. Accessed 21 February 2017.

Vendler, Helen 1995. *The Given and the Made: Recent American Poets* London: Faber; T. S. Eliot Memorial Lectures.

Vogt, Kristiina A. et al. 2010. *Sustainability Unpacked: Food, Energy and Water for Resilient Environments and Societies*. London and Washington, DC: Earthscan.

12

Circles unrounded: sustainability, subject and necessity in Yann Martel's *Life of Pi*

Louise Squire

Yann Martel's Man Booker Prize-winning *Life of Pi* (2002 [2001]) depicts the story of Pi, a boy who finds himself stranded on a lifeboat in the vast Pacific Ocean with a Bengal tiger. Having grown up in the setting of his family's zoo in Pondicherry, Pi is faced with the loss of his family, who – on their way to a new start in Canada – go down with the ship, along with the remaining zoo animals. The central storyline, located in part 2 of the novel, is that of Pi's lifeboat journey, his struggle for survival and his gradual realisation that the wellbeing of the tiger is tied up with his own. Both Pi and the tiger eventually reach the shores of North America, following an extraordinary journey of extreme physical duress and a series of events bordering mainly on the fantastic. As a tall tale with a magical realist feel, the novel is not – or at least is not directly – about sustainability. Indeed, to view it as such raises difficulties, for example in the way it locates its resolution in the Global North. Nonetheless, *Life of Pi* can be read as replicating and playing out some of sustainability's tensions as a concept with complex implications. As such, it ruminates upon challenges faced as sustainability infiltrates contemporary and popular consciousness, taking shape as a present day concern.

Life of Pi has already been discussed in terms of Pi's changing responses to the tiger, unexpectedly named Richard Parker, and the implications of these changes with regard to the nonhuman world and to human–nonhuman dualities (for example, see Huggan and Tiffin 2010; McFarland 2014; Westling 2014). My concern in this chapter is more particularly with the effects of the novel's testing of the human subject horizon – its possibilities and limits – in connection to present-day environmental concerns. This is seen in the way the novel places Pi, as first-person narrator, in juxtaposition with the necessities of a sustainable world, by which, for the purposes of this essay, I mean a world whose ecological capacity to continue to support human life is safeguarded (ecological sustainability). I identify such a world as represented – if figuratively so – in the novel's proleptic account of Pi's eventual establishment of ongoing family life and living in Canada. But this projection of a future world and its wellbeing raises the question of it being possible to account at all for the real beyond grasp, and not just that of some imagined future. Sustainability assumes, moreover, not just such accounting, but – as delineated in the Brundtland definition of sustainable development – the embedding of such accounting in the actions of the present.[1] That is, the wellbeing of future generations is recognised as relying upon our actions going forward today. It is this aspect of sustainability – its entrance into paradigms of the (collective or individual) subject in the present – that is of particular interest in this essay. What does it mean to align our immanent actions with such ungraspable realities as the needs of external others, human or nonhuman, present or future? How, moreover, do we respond to the interruption imposed by these needs upon present paradigms as might otherwise unfold – and at what cost? In order to illustrate the novel's explorations of these difficulties, I read *Life of Pi* alongside a parallel tension in contemporary theory between the phenomenological and the speculative real, about which I say more shortly. Accordingly, I read Pi's personal struggle, conveyed through his first-person narration, as an assertion of the phenomenological, whereby the subject is inescapably positioned within a horizon; and I equate the novel's depiction of the nonhuman realm and its projection of a sustainable world with speculative realism's goal of establishing the outside of thought.

Such an approach to the novel draws attention to the poles between which its narrative operates. In performing the tension between the phenomenological and the speculative real, the novel alludes to such tensions in sustainability as those between its global and local dynamics and between its weaker and stronger forms. The need for ecocritics to pay attention to issues of scale, such as that of the local and global, has been stated before (Clark 2011; Heise 2008; Keller 2012; Trexler and Johns-Putra 2011). The novel handles such issues by directly inserting the object (the futural vision) within the frame of the present, bringing disparate poles into provocative proximity. Thus, in part 1 of the novel we encounter both Pi's account of his childhood in Pondicherry and, interspersed with this, the author's proleptic account of meeting Pi in Canada, years after the voyage, where Pi is by now raising a family of his own. While Pi's first-person narration focalises the addressing of his immediate survival needs, the reader is nonetheless informed – even before Pi's journey at sea begins – that 'this story has a happy ending' (Martel 2002: 93). As such, the novel effectively bridges the 'gap' (see Kate Rigby's essay in this collection) between the unsustainable paradigms of the present and the envisaged ideal of a sustainable world. This is in contrast to the post-apocalyptic envisaging in much environmental crisis fiction of the world we have failed to sustain. Yet it does so by enacting a forced closure, the costs of which are either humorously dismissed or given over to a transcendent notion of change – although they retain a certain, stark presence.

The novel performs this closure through a series of metaphors. Pi's lifeboat journey becomes a metaphor for humanity in the face of current challenges, the tiger a metaphor for the nonhuman world that must be somehow accommodated, and the novel's 'happy ending' a metaphor for the temporal goal of a sustainable world – however conceived. Within the world of the novel, this structural juxtaposition places the character Pi under extreme personal duress, since the predetermined outcome relies on his transforming in two distinct ways: by vastly multiplying his resourcefulness, and by acquiring a new conception of 'other' – both of which turn out to be necessary to his survival. Through imposing such a dynamic, placing Pi under a fiscal austerity of sorts, the novel explores not just the challenges faced by collective humanity but also the difficulties and costs to which such challenges give rise. Through Pi's first-person narration, the novel foregrounds the impacts of perspective, the impossibilities of (absolute) knowing, the possibilities of encounter, and the ways in which intention and necessity can put us at odds with ourselves. As the novel progresses, Pi's subject limitations are gradually overcome, although not so much by choice as by necessity. Pi's survival and that of the tiger turn out to be inseparable, his battle with the tiger – as friend and foe – to be equally a battle with himself. As the novel explores this subject–world dynamic, it replicates an all-important question of the era: how might humanity accommodate that which exceeds it – in theory, in thought and in actuality?

I begin the chapter with a brief discussion of sustainability before going on to relate sustainability to theoretical tensions between phenomenology and the speculative real. I then turn to considering *Life of Pi*'s emphasis on a human-centred stance, alongside its apparent recalibrating of the subject horizon as a sustainable world is engendered.

Sustainability and the human project

A number of sustainability's tensions and paradoxes and their nuances have been teased out across the essays in this collection. This final essay considers sustainability from the perspective of opacity itself. That is, it addresses the issue that sustainability is premised upon projected notions that are variously indistinct or beyond perceptive grasp. These include the very idea of a sustainable world, as well as the endless parts of that which such a world might sustain. In practical terms, of course, sustainability operates at such graspable scales as that of the institution and its policies. Indeed, it (loosely) offers a framework for doing so. Yet it also anchors intentionality to the broader notion of a changed world in which resources are not depleted, species not lost, and so on. While elements of this vision, such as the safeguarding of a certain species, might be mapped out as discrete aims, sustainability often seems not to give full weight to that which it assumes. It is ultimately a slippery term, fraught – as essays in this collection variously observe – with ambiguity and paradox. Sustainability, one might say, exceeds itself as a concept or sign.

As such, where it has as its goal, even if partially drawn, the ecological health of the planet and its parts, human and nonhuman, sustainability reaches – like other environmentalisms – for that which is beyond grasp. Yet it is also quite specifically calibrated to a concern with human wellbeing, actively implicating the human subject, individual or collective. That is, it adopts a human-centred, or anthropocentric, stance. While the emphasis in contemporary ecocritical and other scholarship is often on overcoming the limitations of human-centred thinking, seen for example in much post-humanist and new materialist thought, sustainability is located firmly within human practice and politics – a feature that might be viewed as its weakness or its strength (or both). Outcomes include, on the one hand, its widespread proliferation across the socio-political scene and, on the other, its tendency to lose sight of the environmental matters with which it is concerned. As Stacy Alaimo puts it, sustainability can evoke 'an environmentalism without an environment, an ecology devoid of living creatures other than human beings', whereby 'the lively world is reduced to the material for meeting "needs"' (2012: 562). Alaimo is not pointing here to the loss of other creatures, it seems, but to the kind of instrumental reasoning that perceives them from the subjective limits of human needs-processing.

The tension between sustainability's embeddedness in the human sphere, and its striving to address that which exceeds it, provides one facet of its slipperiness. This is seen, for example, in distinctions between its weak (reformist) and strong (transformational) forms – neither of which escapes the troublesome issue of anthropocentrism. How much change is required and of what kind? In anthropocentric reformism, our unsustainable practices are not blamed on a failure to embrace the nonhuman world, but on what David Kronlid describes as a 'specific kind of anthropocentrism; a shortsighted, ecologically greedy, and ecologically uninformed anthropocentrism that does not take into consideration the ineffective use of nature's resources' (cited in Kronlid et al. 2003: 643). In other words, it is not that we have not recognised the intrinsic value of the nonhuman (deep ecology), but that we have failed to notice, or stupidly ignored, the damage our practices cause to that upon which our survival relies (shallow ecology). The emphasis going forward becomes one of reforming practices as they are now, rather than on addressing overall planetary wellbeing, which would necessitate the more radical changes of a transformational approach.

Transformational sustainability places greater emphasis on change in relation to the global vision of a sustainable world, even if this change is incrementally conceived (Brand 2016; Clifton 2010). One might consider this in relation to Lynn Keller's remark that 'thinking towards sustainability' requires 'local practices' to 'be reconceptualised within planetary dynamics' (2012: 584). While weak sustainability also recognises such dynamics, a transformational response involves a 'fundamental system change' that is non-linear and does not prioritise any particular temporal or spatial scale (Brand 2016: 24). It therefore enacts a radical shift whereby planetary considerations might unsettle established values and practices (Nalau and Handmer 2015, cited in Brand 2016). That such 'planetary dynamics' have an aesthetic dimension points to the value of a literary response to sustainability. Indeed, *Life of Pi* provides a narrative space in which such reconceptualisation is imaginatively explored. It is this that gives rise to the duress placed upon the character Pi within the world of the novel.

The novel poses big questions about the role of the human in relation to change. As Pi (humanity) is set, metaphorically, within a set of global dynamics, he embodies the subjective position from which change is negotiated or might unfold. From here, he is impacted by the circumstances in which he finds himself. The need to account for the world beyond immediate perception is thus depicted as based on necessity, not ethics, intervening in notions of sustainability as shallow or deep while working from a human starting point. One might therefore view the novel's explorations of sustainability's anthropocentric mode, not so much as a means to prioritise human need (although this continues to be addressed), but as a means to frame the problem as a human problem.

The challenges of perception – phenomenology and the speculative real

In playing out the challenges of envisaging change in a contemporary world, Pi demonstrates the difficulties of the partial nature of perception and the limitations of seeing beyond ourselves, especially at the scales and the levels of complexity required. Such limitations are increasingly problematic in the globalised era of environmental crisis. For example, as Timothy Clark – discussing the problems posed in the present to phenomenology – puts it: where, 'especially in the Anthropocene, does "my environment" end', given that '[s]omeone living a high-carbon lifestyle in New York or the Scottish Highlands is already lurking as a destructive interloper on the floodplains of Bangladesh?' (2014: 284–8). But how might such limits be addressed with regard to our human actions and their possible effects?

One advantage of fiction is its capacity to expose the difficulties of envisioning while effectively taking us beyond such limitations. Through opening up disparate spatial and temporal elements fiction can illustrate the partial nature of perception whilst drawing close that which is otherwise inaccessible for interrogation. Such effects appear in specific ways in *Life of Pi*. Its narrative rarely moves away from Pi's first-person focalising of events but creates a fluctuating effect as he narrates from differing times and locations. However, the juxtaposition of his immediate challenges with his 'happy ending' to come sets up a dynamic that clearly exceeds him. This technique of fixing an (externally located) outcome opens up a space by which the challenges of reaching such an outcome are explored. The pressure under which this places the novel's metaphors (Pi, the tiger, the 'happy ending') blurs the borders of reality and fantasy to the extent that the realism the novel largely upholds is rendered precarious, potentially only further emphasising the difficulties of seeing beyond ourselves. Nonetheless, the narrative's centralising of Pi's first-person narration, interspersed with his interpellation by the world that exceeds him, amplifies and thus makes available for exploration the challenges of achieving a sustainable world.

In the analysis that follows, I consider the novel's interrogations of sustainability's challenges through its narrative techniques alongside a related moment in contemporary theory. Specifically, I refer to a tension discernible between the more established theoretics of phenomenology and emergent ideas in speculative realism – a mode of theorising originating at a symposium at Goldsmiths, University of London, in 2007. Tom Sparrow refers to this tension as 'the end of phenomenology' (2014), while Clark remarks: 'all that is most challenging in the twenty-first century about the environmental crisis – politically, sociologically, and philosophically – can be gauged to the degree to which it challenges or even eludes altogether a phenomenological approach' (2014: 284). If by phenomenology we mean the study of the experiential mode of the subject, or the 'structures of consciousness as experienced from the first-person point of view' (Woodruff-Smith 2003, updated 2013), then its 'end' might be found in the emergence of the speculative real. The broad aim of speculative realism might be understood as the desire to 'reconnect philosophy to the "great outside" of the inhuman and ultimately contingent world' (Padui 2011: 90–1). Accordingly, one of its objectives is to overcome that difficulty of post-Kantian thought, defined by Quentin Meillassoux as 'correlationism', whereby: 'All we ever engage with is what is given-to-thought, never an entity subsisting by itself' (2012: 36).

Intriguingly, both Sparrow, who attacks phenomenology from a range of perspectives, and Clark, who notes the weaknesses of its inherited forms, retain some interest in its possible future. For Sparrow, this would mean a return to Hegel's 'absolute idealism', whereby 'Phenomenology could transform idealism into a new variant of speculative realism, and thereby forge a subterranean portal to the things themselves' (2014: 189). Clark considers 'a new ecophenomenology', referring to the work of David Wood, whose aim is to develop 'a middle ground between phenomenology and naturalism, between intentionality and causality' (Wood n.d.: 78). Wood seeks to elucidate, '*a model of the whole as something that will inevitably escape our model of it*' (2003: 8; author's italics). He invokes what he calls the 'plexity' of time (n.d.: 3), through which he demonstrates the incompleteness of our grasp of reality within the temporal plane, thereby countering the 'premature closure' of phenomenology's inherited forms (Clark 2014: 288). An example Wood gives is that of a tree outside our window, which we regularly see, and yet, he suggests, the *life* of the tree, or the living (and temporal) tree, of which we 'glimpse only a limb here, a trunk there', is effectively invisible to us (Wood n.d.: 5). According to Clark, a new ecophenomenology (such as Wood's) would recognise both the complexity of phenomena *and* the way such phenomena can so easily be hidden from view (2014: 288). It would remind us that 'the whole' is dependent on the continuing coordination of parts that have, albeit residual, individual interests' (Wood 2003: 226–7, cited in Clark 2014: 288); thus it would deal *both* with the idiosyncratic interior of the experiential subject *and* with its partial relationality with regard to an exterior or whole.

Ecophenomenology, it appears, makes room for, while not accounting (*as such*) for, the real beyond the phenomenological subject. Such a 'real' is the theoretical goal of speculative realism. Levi Byrant, for example, develops a variant of object ontology[2] by 'bracketing' our preoccupation with access to beings or reality (correlationism) in order to focus instead on 'difference' as 'a matter of the "things themselves", not *our relation* to them' (2011: 262, 267). 'If something makes a difference', he states, 'then it is, full stop' (Byrant 2011: 268). His emphasis, here, is on differences that make a difference, regardless of whether that difference impacts on us. Nonetheless, this idea usefully theorises the way circumstances press upon Pi by bracketing (temporarily) his first-person narration. As such, it provides a means to explicate the way the novel on the one hand says very little about the real it designates beyond Pi's grasp, yet, on the other, *makes* it make a difference to Pi through structural amplification, generated by the use of narrative juxtapositions.

Pi's first-person narration effectively sets up an (eco)phenomenological exploration of the challenges he faces. He operates from the central point of the correlation, his narration describing the world through his access to it; yet he appears not just limited but enabled through his intentional interrelations with the phenomenal world, which is demonstrated to enact difference. My interest, ultimately, is in the novel's seeming insistence, nonetheless, on the correlation at a time when the challenge of environmental crisis is precisely that of our subject limitations. Why emphasise the human perspective at all when our harms to the world are a result of our failure to see beyond immediate needs? And what does this say about sustainability and its human-centric modes?

Life of Pi: circles and horizons

Turning to Martel's novel, the first thing to remark upon is its title, with its metaphorical use of the mathematical symbol, Pi (π). Important to this is the idea, simply, of a circle, by which the novel denotes the horizon of human perspective – or, indeed, that of any sentient being. In referring to the ratio of a circle's circumference and diameter, Pi also has the curious quality of being constant, whatever circle it is applied to. One might extend this to say that any subject (human or nonhuman) is subject to the limitations of its own horizon – whatever the scope of that horizon. Pi (π) is also an irrational number (it is not expressible precisely as a common fraction) and a transcendent number (meaning that one cannot square a circle). If the novel makes mathematical use of these, I have not deduced it, but it might be said to incorporate 'irrationality' and 'transcendence' as aspects of the human subject. Finally there is the question of the title's meaning in reversal: 'circle of life', which points to the conception of life as cyclical, or the cycle of life and death. Along with these meanings, each of which is only partially drawn, the novel's use of a circle metaphor is also intensified within the narrative itself, through the technique of a telling that is more inclined to be circular than linear. Events are narrated from starting points anywhere on a circumference, or unfold temporally in reverse spirals, starting with effects and working back through causes. While these strategies at times make circles and spirals of the plot, they also generate a security in reception, since it is through this that we know of Pi's future wellbeing – although not yet that of the tiger – while reading of his near-starvation at sea.

The novel uses various means to present Pi as bounded within a horizon of knowing, and for much of the novel – throughout Pi's recounting, in part 1, of his childhood in Pondicherry, and in the earlier stages of his lifeboat experiences in part 2 – his capacity to account for others seems, accordingly, limited. This is conveyed through the novel's layering of Pi's narrative voice, juxtaposing the viewpoints of his earlier and later selves. An example of this is the scene in which Pi gives a defence of zoos. Based on his experiences, Pi's reflections are well-meaning and include several judicious points with regard to creaturely needs, encouraging the animal-savvy reader to credit Pi with insight. As a boy, Pi is already open-minded enough to profess to being Hindu, Christian and Muslim all at once. Yet he also tends towards extreme conclusions. He observes, quite aptly, that we tend to romanticise wilderness: 'People think animals in the wild are "happy" because they are "free"' he remarks, describing this as 'nonsense' (Martel 2002: 15).[3] However, his view that nondomestic animals 'live lives of compulsion and necessity' (16) characterised by high levels of fear and hunger, constant territorial challenge and torment by parasites, overlooks, among other matters, the fact that suffering might occur in wild or domestic situations. Pi's conclusion, that zoos answer survival challenges for animals as houses do for humans, based on the observation that a good zoo provides for a creature's needs, again is partially drawn. As Westling notes, zoos may well play a role in safeguarding individuals or species whose habitats are lost (2014: 129). But Pi takes this further, claiming that animals would choose zoos if given the choice. His position is in the end undermined, not just in the face of bad zoos but in the ironic contextualisation of his own analeptic observation in the same section: that Pondicherry Zoo no longer exists. Pi's father closes the zoo when he becomes unnerved by the political climate of New India. Although some animals are rehomed, most end up interned on the ill-fated ship. Thus, what is ultimately emphasised as far as captive animals are concerned – and still through Pi's first-person narration – is their vulnerability, like that of humans, to the realm of human socio-political instabilities.

This already begins to demonstrate the novel's subtle use of Pi's first-person narration to explore issues of anthropocentrism. Given that the novel goes on – as Sarah E. McFarland (2014: 160–2) observes – to variously 'challenge' Pi's pro-zoo stance, it seems too straightforward to say that this instance is illustrative of an overall anthropocentrism in the novel, as suggested in Phillip Armstrong's reading (2008: 178). The novel does, I would suggest, sustain a certain anthropocentrism beyond this episode; indeed my argument is based on it. However, rather than view this as evidence for some lack in the novel, I see it as a device by which the novel explores, testing the limits and possibilities of, a human-centred perspective. Presented through Pi's first-person narration and complicated by the juxtaposition of his childhood and later hindsight views, this reflection on zoos both demonstrates the premature closure enacted by the phenomenological subject and opens the narrative up to the ways in which our perspectives can shift over space and time. The novel's combining of disparate temporal reference points replicates Wood's claim that we only glimpse moments of the real, never the real in itself. Thus, although he attempts to do so, Pi fails to say what nonhuman animals need, perhaps reflecting that what nonhumans need is in an important sense unknowable to us.

Another example of the novel's setting up and undermining of Pi's phenomenological position – or his horizon of knowing – is through his occasional transition to unreliable narrator. Such slippage is apparent already in Pi's recounting of his memories of Pondicherry, but is seen again, with increasing effect, as the novel goes on. It is largely diminished during his initial time at sea although the novel continues to remind us that, much as Pi is rational, observant and well-adjusted, he is capable of mistakes and misjudgements. For example, having described the tiger climbing aboard the lifeboat, he recounts three days at sea in which no tiger features, other than as the figure of his anxiety. He thinks he must have imagined the tiger, but it turns out to be lying low, suffering from sea-sickness under a tarpaulin. This depicts, not so much a failure on Pi's part, but the limitations of perception, especially under trying circumstances. While it seems unlikely that one could spend three days on a lifeboat with a Bengal tiger and not be sure of its presence, it is not beyond the realms of possibility.

As the novel goes on, its use of the unreliable narrator intensifies. Thus, having spent many months at sea, Pi recounts an encounter with a floating carnivorous island, populated by nothing but meerkats. Later still, weakened by exhaustion, he recounts going temporarily blind and entering into a discussion – he thinks – with the tiger, which on regaining his sight turns out to have been an old man in another lifeboat, now eaten by the tiger. In these episodes, a blurring of the boundaries of reality occurs, yet the novel maintains sufficient sense of the real that the reader – rather than drift into fantasy – becomes conscious of her own limitations. We are reminded that the world is not always as we perceive it to be, that we are susceptible to imperception, irrationality and even hallucination. This use of the unreliable narrator to render fallible the human subject is finally driven home in the third section of the novel, when, having reached North America, Pi is questioned by officials who do not believe his story. Obligingly, he offers a different story in which the passengers on the lifeboat were really his mother, the ship's cook and a sailor, and he asks the officials to choose whichever story they prefer. This has the dramatic effect of leaving the reader with no way of telling which story is supposed to be the real one. The novel clearly intends no resolution here, and so the reader, as subject, finds themselves in the last instance to be fully undermined.

In its representations of the subject as caught in its own horizon of knowing, the novel demonstrates two of Clark's points with regard to Wood's ecophenomenology. Firstly, it demonstrates the subject's partial relationality with regard to the whole; thus Pi, as with the view of a tree outside a window, responds to life's realities as they appear before him, even as his access to them is always in glimpses. Secondly, it takes account of the idiosyncrasies of the experiential subject. This is important in that achieving a sustainable world inescapably depends upon people – whether those in power or the general populace; yet people, including those who prioritise sustainability, are not necessarily dependable. This difficulty is usefully theorised in Adeline Johns-Putra's new materialist critique of environmental care ethics, in which she suggests that care, rather than being the means by which agency occurs, is 'itself agential'; thus, she argues, 'the agency of caring is contingent on the level and quality of caring,' and caring is in turn 'always contextualised' (2013: 126, 133). One might want humanity to care (for the planet and its creatures), and want this care to be effectively deployed, but care is always to an extent governed by (often conflicting) experiences and by responses to these experiences.

Pi's journey

Pi's subjecthood, accordingly, is not static throughout the novel. Circumstances press upon him and change him as he is cast adrift in the Pacific Ocean. This 'pressing upon' is reminiscent of evolving ideas in the present era, in which unquestioned trajectories of human progress are 'interrupted' by the advent of an external crisis of environment. Pi's journey of survival begins with the sinking of the ship, aboard which are his family and many of the zoo animals. This episode sets up certain metaphors by which we might read the novel's treatment of sustainability. The sinking of the ship might denote the fall of the social world – a collapse commonly depicted in environmental crisis fiction, based on society's failure to enact the changes required. (For discussions on societal collapse, see Dana Phillips' essay in this book.) For Pi, the sinking of the ship is 'as unbelievable as the moon catching fire' (Martel 2002: 103), perhaps reminding us of the difficulty of imagining a world without capitalism (Jameson 2003). Pi sees his life as entrusted to 'the officers of our destiny' (Martel 2002: 104), the ship's crew whom he remains convinced have a handle on the situation (even as the ship fills with water and lists to one side), and whom he believes will respond accordingly. But, perhaps in keeping with leadership issues of today's world, the officers instead display an irresponsible denial of the situation, unceremoniously throwing Pi into a lifeboat occupied by a hyena. Only later does Pi suspect their intention to have been to free up the lifeboat for themselves.

Having dispensed with the social order – a process to which Pi's own family and the many interned zoo animals are tragically sacrificed – the novel transfers Pi's journey (to a sustainable world) from the domain of the ship to that of the lifeboat. Here, the lifeboat, as Westling remarks, becomes an 'elegant material pun on what earth is for us' (2014: 126). We might understand this as conjuring up a previously un-grasped external real (in this case the Earth, the environment or nonhuman world) within the orbit of Pi's perception. In depicting the lifeboat, the novel also alludes, as noted by Huggan and Tiffin, to survival narratives such as *Robinson Crusoe* (2010: 174). One might also view the lifeboat in relation to Garrett Hardin's (1974) 'lifeboat ethics', disturbingly subtitled 'the case against helping the poor', in which he argues that the planet can only support so many lives. Each of these allusions pertains to resources: to our various uses of them, our actions in the face of their finiteness, and so on, thus evoking a sustainability theme. In dealing directly with the situation, Pi must contend with both the situational limitations and his own – a process that significantly challenges his sense of being and his views on life. In metaphorical terms, he is forced to confront the gap between sustainability as local project and sustainability in the overarching sense of a sustainable world.

There are two notable aspects to Pi's changing responses at sea. One is the shift in his relations with the tiger, the other his growing awareness of the physical world, as represented by lifeboat and ocean. Each of these in some way alters his perspective. At first, Pi's responses to the tiger are, quite understandably, dramatic and extreme. Urging him to swim to the safety of the lifeboat, Pi suddenly realises the implications. As the tiger clambers aboard, Pi jumps off into the sea, unable to imagine their mutual survival. For some time after that, he gives witness to the battle that rumbles on between the various nonhuman occupants of the lifeboat: the hyena, a zebra with a broken leg, an orangutan and the tiger. The battle is physical, since the hyena kills the zebra and the orangutan, only to be killed by the tiger; and it is psychological, since the terrified Pi must use his mental resources to generate physical resources, building a raft upon which to drift some distance from the lifeboat with its resident tiger. Pi battles too with his own sensibilities: as a vegetarian and a pacifist, he must negotiate his horror at the task of dispatching sealife to survive. Yet each of these battles – headed by his fear of the tiger – is overshadowed by his most basic need: for water. 'With a tiger on board', he remarks, 'my life was over. That being settled, why not do something about my parched throat?' (Martel 2002: 135). Necessity is thus distilled to ultimate necessity.

The gradual changes in Pi's responses to the tiger have been variously observed. Huggan and Tiffin, referring to Pi's various dominance displays, note that he learns to control the tiger 'by acknowledging and re-inhabiting his own animality, not by divesting himself of it' (2010: 172). This is illustrative of the novel's gradual blurring of human–nonhuman boundaries. Importantly, an agential reversal is also at work; thus, as McFarland notes, 'although Pi coaches Richard Parker to respect his territory on the lifeboat, Richard Parker also trains Pi to read his signals' (2014: 158). A key instance of this occurs when Pi recognises the tiger to be communicating with him in a certain way. 'Prusten', Pi explains, is 'the quietest of tiger calls, a puff through the nose to express friendliness and harmless intentions'

(Martel 2002: 163). It is following this that Pi concludes: 'it was not a question of him or me, but of him *and* me. We were', he states, 'literally and figuratively, in the same boat' (164). Consequently, of his list of seven ways to deal with the tiger, most of which involve the tiger's necessary death, Pi decides on the seventh, which is '*Keep Him Alive*' (166). This is a crucial moment in the novel, since it signals the recognition that human survival (that of Pi) is ensured, not through the instrumental reasoning of destruction, but through co-existence with the nonhuman realm (represented by the tiger). This transition is verified in Pi's subsequent recognition that the tiger, literally, has kept him alive. His eventual affection for him is most poignantly portrayed when they reach North America and the tiger jumps ashore, heading into the trees without looking back. Devastated that the tiger does not somehow 'conclude' their 'relationship' (284), Pi wishes he had found some way to thank him. Instead, he is forced to accept that his care for the tiger must remain unacknowledged. Here, human accommodation of the nonhuman appears as based on necessity, not attachment, and accordingly is beyond conditionality.

Pi's journey at sea also involves his response to the physical world. This has two aspects: Pi's physical survival and the changing dimensions of his mental landscape. An example of this is his belated discovery of the boat's stores and his learning to use its resources while accounting both for his needs and those of the tiger. The stores contain any number of useful items from fishing gear to solar stills (which turn seawater into drinking water), as well as sufficient food and water for ninety-three days. As he grows in resourcefulness, Pi learns to make full use not just of the boat's stores but of anything available, enabling him to sustain himself and the tiger when rations run out. This involves his constant discovery of what is already before him. Describing the physical features of the boat, he remarks:

> I did not grasp all these details – and many more – right away. They came to my notice with time and as a result of necessity. I would be in the direst of dire straits, facing a bleak future, when some small thing, some detail, would transform itself and appear in my mind in a new light. (Martel 2002: 139)

Possible solutions are already present, the novel suggests; it is the recognition of necessity that brings them forward into view; what limits us is a failure to grasp necessity.

Although Pi's circumstances are minimal and dire, the intensity of his hunger also brings to life his visual imagination. He discovers awe in simple things, such as the tiger's agility in dispatching a flying fish, or the drama of a storm. Increasingly, Pi's 'noticing' involves the world as it goes on about him, the lives of others also striving to survive. Hearing noises below the boat, he realises that, 'The battle for life was taking place there too' (119). Looking overboard, he remarks, 'With just one glance I discovered that the sea is a city. Just below me, all around, unsuspected by me, were highways, boulevards, streets and roundabouts bustling with submarine traffic' (175). Even the underside of the lifeboat becomes a 'host to a multitude of sea life' (197). But for all his new appreciation, and despite his evolving attention to improvements of the boat and survival arrangements, Pi feels he has descended 'to a level of savagery' he 'never imagined possible' (197). After some time at sea he is able to dispatch a turtle, drink its blood while still warm, and make use of every atom of it as resource without flinching. If, as Huggan and Tiffin (2010: 172) note, he reinhabits his own animality, this animality is also normalised as a mode of being; thus sentiment, as a guide for living, is again overruled by necessity.

Perspectives: subject and world

Despite undermining it at times, the novel's emphasis on the phenomenological is more or less sustained, allowing the reader to witness Pi's sense of what is before him and his responses to the ways in which circumstance press upon him. Late on in part 2, this emphasis is given a direct voice. Pi relates:

> To be a castaway is to be a point perpetually at the centre of a circle. However much things may appear to change – the sea may shift from whisper to rage, the sky might go from fresh blue to blinding white to darkest black – the geometry never changes. Your gaze is always a radius. The circumference is ever great. (Martel 2002: 216)

This reference to the novel's metaphorical use of Pi (π) is complemented here with the point that life is not constant. This is developed in Pi's (preceding) proclamation: 'There were many skies', he states, continuing: 'The sky was a featureless milky haze. The sky was a density of dark and blustery rainclouds that passed by without delivering rain. The sky was painted with a small number of flat clouds', and so on, for over half a page (215). Next, Pi observes: 'there were many seas'; thus: 'The sea roared like a tiger. The sea whispered in your ear like a friend telling you secrets. The sea clinked like small change in your pockets', and so on. There were also 'all the winds' and 'all the nights and all the moons' (215). In other words, located perpetually within a horizon, we cannot grasp the world in any singular sense; it brings itself anew from moment to moment, even as language categorises its parts ('the sky', 'the sea', 'the wind'). In Wood's terms, we encounter it in glimpses.

Similarly, the challenge of sustaining is always immediate, always idiosyncratic to circumstance, to location, to the moment, to the person.

'Sustainability', viewed at close hand, exceeds itself as a concept or sign. Yet, Pi's personal quest for survival is distilled, intense and singular, arising in the inexorable power of hunger and thirst. It is brought to actuality, nonetheless, through his acknowledgement of his partial relationality with regard to the world beyond him. Late on in Pi's lifeboat journey, he juxtaposes these conflicting points, emphasising them: 'I noticed ... that my suffering was taking place in a grand setting. I saw my suffering for what it was, finite and insignificant, and I was still'; next he follows this with the parenthesised, '(No! No! No! My suffering does matter. I want to live ... Life is a peephole, a single tiny entry onto vastness – how can I not dwell on this brief, cramped view I have of things? This peephole is all I've got!)' (177). Pi's inner battle replicates the internal conflicts of sustainability with which humanity must somehow contend. While this might mean that in dealing with immediate circumstance the larger picture must be accommodated, the novel seems to suggest more strongly that accommodating the larger picture is itself the means by which immediate circumstance must be tackled – at least to be effective in global terms.

The novel also undermines Pi's phenomenological positioning, at moments of his unreliability as narrator, and especially in the final chapter as his story is detached from its phenomenological moorings. (Which story is 'real'?) This does not altogether dislodge Pi from his first-person view of his circumstances, but rather makes absolute its incompleteness.

Sparrow, amid his lengthy discussions on the end of phenomenology and the advent of the speculative real, at one point makes an intriguing remark:

> To escape correlationism ... it is not necessary to position oneself at some objective vantage point outside the correlation, which is impossible. It is also not necessary to pass through the correlation, as does Meillassoux. What is necessary is to record the "genetic" movement of and on this world, unconditionally. (Sparrow 2014: 151)

Sparrow's remark, here, perhaps provides the closest indicator of the difficulty the novel tackles as it evokes the tension between phenomenology and the speculative real. If we always, inescapably, operate from within our own vantage point, how can we – or, indeed, *can we*, in literature or in our discursive lives – account for that which exceeds us? But what is meant, exactly, by this *unconditional* 'recording' of a '"genetic" movement of and on this world', and can it have any literary bearing?

Sparrow's point appears to reflect the pivotal distinction between Wood's ecophenomenology and Byrant's realism – a distinction constantly at work in Martel's novel. On the one hand, Pi is at no point removed from the correlation. On the other, the novel intersperses his first-person narration with situational effects, many of which function as figments of the real that 'make a difference' (in Byrant's sense) to him, made possible through the novel's extensive use of metaphor. At the same time, the temporally fractured nature of Pi's first-person narration, such that at times he conveys the immediate and at times speaks from hindsight, can be understood in terms of Wood's 'plexity of time'. Pi encounters the world incompletely, through momentary glimpses, which the novel also juxtaposes to demonstrate his changing responses (Wood n.d.: 3–4). While Pi's position within the correlation is clearly sustained, the incompleteness of this positioning is also rendered absolute.

What, then, are we to make of *Life of Pi*'s explorations of subject and world? As far as the distinction between the phenomenological and the real is concerned, the novel seems to make quite plain its incapacity to represent such a real, while insisting that the real is nonetheless in some unconditional sense, there. It therefore challenges its readers by positioning us in dialogue with the unconditionality of that which exceeds us, although not with that which exceeds us *per se*. With regard to what this says about sustainability, the novel establishes the real world as existing, while insisting on our inability to grasp it in full. Such a world becomes both the subject of our ethical response and the substance of necessity. Thus, Pi's (humanity's) survival relies on his accommodating the tiger (the nonhuman world), but also on his dispatching of sealife. If the first of these engages his ethical response, the second disengages it, drawing attention to sustainability's difficulties. What are sustainability's ethics? We must be clear, the novel suggests: they are based on the will to survive.

Life of Pi does carry problematic undertones in terms of what it says about sustainability. Written by a (male) French-speaking, white Canadian, it elects as its protagonist an Indian boy who hails from a once-French colonial settlement in India, whose life falls apart, who overcomes all odds, and whose 'happy ending' materialises in North America. Pi's journey is at the cost of appalling loss – his parents and brother, his nonhuman companions, his home – and he undergoes near-starvation at sea. These events are structurally linked to the novel's global outcome, which Pi's traumatic transformation achieves. Events read from this perspective depict the novel as evoking a so-called First-World assertion of sustainability's demands.

Yet the novel might also be read as playing out and exploring some of sustainability's conflicting elements. As it negotiates the local in terms of planetary dynamics, it reflects on the competing paradigms of contemporary reality. Pi's exposition of many winds and many oceans alludes to life's many stories and many realities, and to the ways these become tangible in glimpses, if at all. At the same time, the novel insists that, in the face of reality, something must give. As such, Pi's suffering within the frame of his transformation symbolises the necessity of a radical response, yet also foregrounds its costs. While these ought to be the indulgent lifestyles of those who can afford them, or the political systems that privilege corporate interest over environmental concerns, they might well be the lives/wellbeing of those whose interests slip out of view. Ultimately, in framing sustainability as a human problem, *Life of Pi* renders it tangible as a question that demands a response – the dynamics of which are in constant negotiation.

Notes

1 I refer here to Gro Harlem Brundtland's definition of 'sustainable development', which refers to development that 'meets the needs of the present without compromising the ability of future generations to meet their own needs' (World Commission 1987: 43).
2 Object ontology – which originates in the work of Graham Harman – is, according to Sparrow, a 'fully committed realist metaphysics' that focuses on objects rather than on the human (2014: 114).
3 All references to *Life of Pi* are to this edition.

References

Alaimo, Stacy 2012. 'Sustainable This, Sustainable That: New Materialisms, Posthumanism and Unknown Futures', *PMLA* 127 (3): 558–64.
Armstrong, Phillip 2008. *What Animals Mean in the Fiction of Modernity*. New York: Routledge.
Brand, Ulrich 2016. '"Transformation" as a New Critical Orthodoxy: The Strategic Use of the Term "Transformation" Does Not Prevent Multiple Crises', *GAIA: Ecological Perspectives for Science and Society* 25 (1): 23–7.
Byrant, Levi R. 2011. 'The Ontic Principle: Outline of an Object Oriented Ontology'. In *The Speculative Turn: Continental Materialism and Realism*. Ed. Levi Byrant, Nick Srnicek and Graham Harman. Melbourne: re.press, 261–90.
Clark, Timothy 2011. *The Cambridge Introduction to Literature and the Environment*. Cambridge. Cambridge University Press.
Clark, Timothy 2014. 'Phenomenology'. In *The Oxford Handbook of Ecocriticism*. Ed. Greg Garrard. New York: Oxford University Press, 276–90.
Clifton, Don 2010. 'Representing a Sustainable World: A Typology Approach', *Journal of Sustainable Development* 3 (2): 40–57.
Hardin, Garrett 1974. 'Lifeboat Ethics: The Case Against Helping the Poor', *Psychology Today* 8: 123–6.
Heise, Ursula 2008. *Sense of Place and Sense of Planet: The Environmental Imagination of the Global*. Oxford: Oxford University Press.
Huggan, Graham and Helen Tiffin 2010. *Postcolonial Ecocriticism: Literature, Animals, Environment*. Oxford: Routledge.
Jameson, Fredric 2003. 'Future City', *New Left Review*, 21. http://newleftreview.org/II/21/fredric-jameson-future-city. Accessed 21 February 2017.

Johns-Putra, Adeline 2013. 'Environmental Care Ethics: Towards a New Materialist Critique', *Symplokē* 21 (1–2): 125–35.

Keller, Lynn 2012. 'Beyond Imagining, Imagining Beyond', *PMLA* 127 (3): 579–85.

Kronlid, David et al. 2003. 'Behaviour and the Environment Ethics, Education, and Lifestyle'. In *Environmental Science: Understanding, Protecting and Managing the Environment in the Baltic Sea Region*. Ed. Lars Ryden, Pawel Migula and Magnus Andersson. Uppsala: Baltic University Press, 630–61.

McFarland, Sarah E. 2014. 'Animal Studies, Literary Animals, and Yann Martel's *Life of Pi*'. In *The Cambridge Companion to Literature and the Environment*. Ed. Louise Westling. New York: Cambridge University Press, 152–65.

Martel, Yann 2002 [2001]. *Life of Pi*. Edinburgh: Canongate.

Meillassoux, Quentin 2012. *After Finnitude: An Essay on the Necessity of Contingency*. London: Bloomsbury.

Padui, Raoni 2011. 'Realism, Anti-Realism, and Materialism', *Angelaki: Journal of the Theoretical Humanities* 16 (2): 89–101.

Sparrow, Tom 2014. *The End of Phenomenology: Metaphysics and the New Realism*. Edinburgh: Edinburgh University Press.

Trexler, Adam and Adeline Johns-Putra 2011. 'Climate Change in Literature and Literary Criticism', *Wiley Interdisciplinary Reviews: Climate Change* 2: 185–200.

Westling, Louise 2014. *The Logos of the Living World: Merleau-Ponty, Animals and Language*. New York: Fordham University Press.

Wood, David n.d.. 'What is Eco-Phenomenology?'. Adapted version of chapter in *Eco-Phenomenology: Back to the Earth Itself*. 2003. Ed. Charles S. Brown and Ted Toadvine. Albany: State University of New York Press. http://pages.uoregon.edu/toadvine/. Accessed 21 February 2017.

Woodruff Smith, David 2013. 'Phenomenology'. *Stanford Encyclopedia of Philosophy*. https://plato.stanford.edu/entries/phenomenology/. Accessed 30 March 2017.

World Commission on Environment and Development 1987. *Our Common Future* (Brundtland Report). Oxford: Oxford University Press.

Index

Literary and cinematic works can be found under authors' or directors' names, and 'n.' after a page reference indicates the number of a note on that page

Abecassis, Jack 208
activism xiv, 4, 90, 93–4, 105, 147
Adams, Carol 105
Adorno, Theodor 58, 69, 153
agriculture 18, 21, 25, 28, 56, 150, 154
  agricultural communities 22, 220
  agricultural scientists 53
Alaimo, Stacy 2, 57, 78–9, 88–9, 90, 93, 216–18, 220, 226n.3, 231
ambiance 38–40
Andermahr, Sonya 182–3
Anderson, Sherry Ruth 196
animality 105, 239, 241
animals xv, 18, 57, 60, 62, 64–6, 70, 97–112, 113n.1, 141–2, 146, 156, 185, 205, 219, 228, 235–6, 238
  feed 28
  rights 107–8, 147
animal studies 97
Anthropocene xv, 7, 60, 68, 116, 127–8, 130–4, 232
anthropocentrism xiv, 2, 53, 57–9, 69, 79, 99, 105, 123, 185, 189, 212–13, 215, 217, 219, 230–2, 235–6
  see also human-centredness
anthropogenic climate change *see* climate change
anthropomorphism 63
apocalypse 18, 109, 140, 145, 150, 165, 169
  see also post-apocalypse
Archer, Jayne Elisabeth 2, 6
Arnold, Matthew 153
Asimov, Isaac 161
Atwood, Margaret
  MaddAddam trilogy 7, 139, 143–53

Babbage, Charles 15
Baggesen, Jens 37
Balzac, Honoré de 41, 44
Barad, Karen 61, 78
Barrell, John 22
Bate, Jonathan 62, 69
Bauman, Zygmunt 80, 87, 93
Becker, Christian U. 179
Bekoff, Marc 104
Bell, Lucy 6
Bellamy, Edward 160
Benford, Gregory 161
Benjamin, Walter 153
Bennett, Jane 78–9, 90–1, 94
Bergthaller, Hannes 4, 7, 48, 141
Bermingham, Ann 21
Bernard, Claude 39, 44
Bible 58, 60–1, 122, 147, 186
  see also Old Testament
biodegradability 117–20, 122, 125, 127, 131
biodiversity 29, 52, 56–9, 71n.4, 97–8, 100–1, 106, 112
biopolitics 99, 108
biosemiotics 37, 71n.8
biotechnology *see* technology
Blake, William 14, 29n.1, 121–5
  'Little Black Boy' 121
Blass, Eddie 160
Blomkamp, Neill
  Elysium 127
Bloomfield, Mandy 2

Bloomfield, Robert 19, 29n.6
Boden, Deidre 21
Bohr, Niels 61
Boyer, Dominic 159
Braidotti, Rosi 78, 208
Bresnihan, Patrick 64, 66
Brin, David 161–2
Brooks, Peter 180–1
Brundtland, Gro Harlem 55, 178, 195, 244n.1
Brundtland Report 2, 5, 7, 55, 57, 79, 92, 99, 100, 178–9, 195, 198, 200–1, 208, 229, 244n.1
  see also United Nations
Brunkhorst, Hauke 93
Brunner, John 161
Buell, Lawrence 40, 43
Buse, Katherine 172
Bush, George W. 163
Butler, Andrew M. 160–1
Butler (Bishop), Joseph 15
Byrant, Levi 234, 242–3
Byron (Lord), George Gordon 17–18

Callenbach, Ernest
  Bring Back the Buffalo! 102, 105–10, 112
  Ecotopia 102, 105–6
Calvino, Italo 80
Camden, William 15–17
Campbell, Colin 63
capitalism 55, 61, 69, 76, 107, 112, 116, 141, 161, 163, 166–8, 170, 172, 177, 189, 196, 202, 205–6, 238
  see also consumerism
carbon xiv, 18, 163, 165–71, 174, 175n.1, 180, 222, 232
carbon dioxide 90, 165
Carlyle, Thomas 35, 37, 40
  Sartor Resartus 35
cartoneras see editoriales cartoneras
Cervantes, Miguel de 15
Chakrabarty, Dipesh 144
Chaucer, Geoffrey 29
  Canterbury Tales, The 20, 23–5
Clare, John 6, 29, 54, 62–8
Clark, Timothy 232–4, 237
Clarke, Arthur C. 161
Clarke, I. F. 161
cli-fi 127, 133 *see also* climate change, fiction
climate change 48, 56–8, 61, 71n.4, 78, 127–8, 143–4, 146, 148–50, 159, 162–9, 170–4, 175n.1, 183, 203, 218
  fiction 149
    see also cli-fi
  global warming 52, 57, 156
Cohen, Tom 218, 220
Colebrook, Claire 6–7
Coleridge, Samuel Taylor 35, 40
collapse 7, 97, 106, 139–41, 143–4, 147–9, 152–3, 156, 238
comedy 164, 166, 168, 174
conservation 7, 53–4, 57, 59–60, 68, 97–101, 103–5, 107–9, 113n.3, 191, 196
  of watermills 27
Constable, John 19, 20–3
consumerism 48, 61–3, 80, 87, 89, 106, 112, 151, 154, 177, 180, 196, 202, 205–6
  see also capitalism
Corcoran, Peter Blaze 211–12, 225
Corn Laws 17, 34, 46
correlationism 8, 233–4, 242
Cruz Barros, Áurea de Lima 90–1

d'Angelo, Ana 84
Danz, Harold 107
Darwin, Charles 201
  Darwinism 148, 161
deconstruction 7, 41, 115–16, 119, 122, 128
deep ecology 2, 7, 53, 157, 161, 172, 177, 196, 231–2
  see also ecology; shallow ecology
DeLanda, Manuel 78
Deleuze, Gilles 78, 124, 128–9
de Man, Paul 122–3, 125–9, 131–2, 134
democracy 42, 61, 122–3, 134n.1, 198–9, 201
Democritus 78
Demosthenes 126
de Mul, Jos 207
Derrida, Jacques 54, 60, 117, 119–20, 123–4, 134n.1
Diamond, Jared 140

Dias, Maria 84–5
Dobson, Andrew 33
Doctorow, Cary 161
Donne, John 184–5
Dresner, Simon 5, 178
Dufourmantelle, Anne 54
DuPlessis, Rachel Blau 181
dystopia and dystopianism 7, 143, 163, 207
  see also utopia and utopianism

ecocriticism 1–5, 14, 37, 40–1, 62, 69, 71n.7, 147, 229, 231
ecofeminism 58, 70n.4, 161
  see also feminism
ecological stability *see* stability
ecology xii–xiii, 2–3, 7, 33–43, 47–8, 52–4, 56–62, 65–6, 69, 70n.4, 71n.5, 78, 91, 93, 98, 100, 102, 105–7, 109, 111–12, 118–19, 140, 147, 150, 159, 161–2, 164–5, 167, 169, 186, 200–3, 205, 208, 212, 228, 231
  movements 33, 105, 157, 196
  science xii, 2, 33, 36, 43, 53, 57, 106, 139–42, 161, 200–1
  see also deep ecology; shallow ecology
ecomimesis 40–1
ecophenomenology *see* phenomenology
ecopoetics 54, 62, 67–8, 70, 215
ecosocialism *see* socialism
Edelman, Lee 181–2, 186, 188, 191–2
editoriales cartoneras 7, 76–7, 81, 95
Ehrlich, Paul and Anne 198–9
  Population Bomb, The 200
Elder, John 213, 219, 224
elegy xiii, 25
Eliot, George 29
  Mill on the Floss, The 19–20, 25–6
Eliot, T. S.
  'Tradition and the Individual Talent' 219
  Waste Land, The 220–2, 226n.3
Elkington, John 77
Ellam, Julie 191
Emboava, Andréia Ribeiro 84
Emboava, Peterson 84
enclosure 14, 65–6, 68, 108
Engels, Friedrich 42
Enlightenment 54, 121, 131, 199, 208
environmental humanities 8, 58, 78–9, 115–16, 197
Epicurus 74
Euclid 120, 127
evolution 37, 41, 47, 59, 155–6, 183, 199, 201, 204, 206
  evolutionary theory 201
extinction xii, 7, 58–9, 70, 71n.4, 97–104, 106–7, 110, 112, 112n.1, 146, 148, 197, 201, 208
  see also populations, animal

fantasy 146, 156, 228, 233, 237
feminism 53–4, 181, 185, 191
  see also ecofeminism
Fischer, Joern 3, 52–5, 57–8
Flechtheim, Ossip K. 160
Fletcher, John 18–19
food 18, 29, 36, 61, 93–4, 107, 146, 151, 198, 206, 220, 240
  chain 13–15, 17–19, 22–3, 24–8, 29n.8, 37, 48, 52, 101, 103
  riots 22, 26, 29n.3
  security 28, 56
Foote, Stephanie 2–3, 5
Foreman, Dave 104, 109
Foucault, Michel 99, 132, 199
Fourier, François 33, 44
Friedland, Roger 21
Frost, Robert xii–xv
futures studies and futurology 159–62, 164

Ganteau, Jean-Michel 187
gender 184
  transgender 111, 122
genre 6–7, 84, 127, 143–4, 146, 152, 156, 162
geoengineering 116, 128, 162, 172, 189
Gernsback, Hugo 160
Gibbons, Stella
  Cold Comfort Farm 29
global warming *see* climate change
Goethe, Johann Wolfgang von 35, 37, 47–8
Goncourt, Edmund and Jules de 41
Gore, Al 163, 167
Gould, Peter 42

Graham, Jorie
Sea Change 7, 211–26
Gray, Thomas xii
Griffiths, Matthew 7
Griggs, Dave 55–6
Grober, Ulrich 34–7, 40–3, 45, 47
Guattari, Félix 78, 124, 128–9
Guha, Ramachandra 90

habitat 56, 58, 60, 65, 68, 70, 97, 99–102, 101, 104, 142–3, 173, 236
Haeckel, Ernst 36
Halberstam, Jack 113n.6
Haraway, Donna 185
Hardin, Garrett 239
Hardy, Thomas 19, 29
Harrow, Susan 44
Hawken, Paul 107
Head, Dominic 41
Hegel, G. W. F. 125, 233
Heidegger, Martin 67, 69
Heinlein, Robert A. 161
Heisenberg, Werner 61
Herder, Johann Gottfried 35, 37, 48
Hertz, Neil 125–6
Hesiod 17
heteronormativity 105–6, 111, 181, 184–5, 191
see also queering and queerness
Higgins, Jack 21
historical fiction 7, 143–4, 146–7, 150–1, 156, 162
Hobbes, Thomas 17
Holling, C. S. 141–2
Homer
Odyssey, The xiii
Hopkins, Gerard Manley 35
Horkheimer, Max 58, 69
Hornaday, William 103
Houellebecq, Michel
Possibility of an Island, The 7, 197, 202–9
Hubbard, Tasha 110
Huggan, Graham 239, 241
human-centredness 53, 105, 230–1
see also anthropocentrism
Humboldt, Alexander von 35–6
Hurricane Katrina 78, 172

Husserl, Edmund 119, 124–5
Huxley, Aldous 152

industrialisation 14, 18, 56, 66–7, 78, 167
farming 46
production 42, 119, 150, 157
society 42, 63, 68, 107, 146, 157
Irigaray, Luce 131
irony xiv–xv, 63, 67, 88, 105, 107, 121, 149, 151, 155, 182, 206, 236

Jameson, Fredric 152
Jarvis, Matthew 2
Johns-Putra, Adeline 2, 7, 71n.4, 163, 238
jouissance 5, 181–2
Joyce, James 124–5
Juventino, Elizângela 84

Kant, Immanuel 125, 233
Kaplan, Steven S. 14
Katoppo, Aristides 92–4
Keats, John xii, 17, 42
Keller, Lynn 2, 6, 232
Kermode, Frank 145, 180–2, 192, 193n.1
Kerridge, Richard 62–3
Kilgore, Douglas DeWitt 164, 168
King, Thomas 110
Klein, Naomi 127
Kroeber, Karl 21–2
Kronlid, David 231
Kurt, Hildegard 3
Kyoto Protocol 196

LaDuke, Winona 109
Langland, William 29
last animals *see* extinction; populations, animal
Latour, Bruno 115–16
Lem, Stanislaw 161
LeMenager, Stephanie 2–3, 5
Levin, Harry 15
liberalism 36–7, 39, 182, 201
liberal humanism 127
neoliberalism 76, 181, 190
Lifestyle of Health and Sustainability (LOHAS) 196–7, 206
Linhart, Danièle 87

Linnaeus, Carl 35–6
LOHAS *see* Lifestyle of Health and Sustainability
Longinus 125–6
Lovins, Amory and L. Hunter 107
Luckhurst, Roger 162–4

McCarthy, Cormac 151
McCulloch, Fiona 191
McFarland, Sarah E. 236, 239
MacKinnon, J. B. 98, 100
magical realism 228
  see also realism
Malthusianism 43, 49, 197–202
  see also populations, human
Manning, Adrian 57
Marcuse, Herbert 39
Markley, Robert 168
Marris, Emma 104
Marsh, George 33
Martel, Yann
  Life of Pi, The 7–8, 222, 228–30, 232–44
Martin, Paul 104
Martinez, Joan Alier 90
Marx, Karl 42
Marxism 42, 69
materialism and materiality 61–3, 66, 78, 88, 116, 118, 121–5, 128–31, 215
  see also new materialism
Mathews, Freya 54, 59–64, 66, 70, 71n.5
Matisse, Henri 223
Meadows, Donella 198
  Limits to Growth, The 200
Medina, Martin 81
Medovoi, Leerom 211–13, 215–16, 221, 225
Meillassoux, Quentin 233
Mentz, Steve 56–7
Merchant, Carolyn 33, 58
metanarrative 144, 211–12, 214–15, 217, 225
metaphor 18–19, 23, 25, 42, 45, 54, 132–3, 182, 230, 232–3, 235, 238–9, 241, 243
Meza, Aurelio 82
Middleton, John 22
Miles, Ian 160–1
milieu 37–41, 44, 118, 125–7, 129
Mill, John Stuart 33, 40, 47
Miller, Nancy K. 181
Milner, Andrew 162
Mocarzel, Eduardo 94
modernity 36, 43, 46, 48, 66–7, 69, 80, 103, 157
Monbiot, George 104
Monkmon, Kent 110
Morley, David 6, 54, 68–70, 71n.12
  Gypsy and the Poet, The 68–70
Morris, William 6, 33, 43, 47
  News from Nowhere 34, 42–4
Morton, Timothy 40–1
Muñoz, José Esteban 188
Muskett, Paul 23
Myers, Norman 101

nationalism 44, 98, 103–4, 108–9, 111–12
  national boundaries 174, 201
  see also transnationalism
Nazism 39
Nelson, Brian 44
new materialism 61, 71n.4, 78–80, 90, 130, 231, 238
  see also materialism and materiality
Nietzsche, Friedrich 128, 204
Nolan, Chris
  Interstellar 127, 133

object ontology 234, 244n.2
O'Grady, John P. 2
oil 26, 159, 167, 174
  peak 141, 159
Old Testament 18
  see also Bible
Onega, Susana 182, 191
Oppermann, Serpil 41
O'Riordan, Timothy 179
Orwell, George 152
Otto, Eric 160–1

Paddock, William and Paul 198
Pak, Chris 7
Parham, John 2, 6
Parmée, Douglas 44
Parry, Anne and Andy 28
pastoral xiii, 65, 163, 167, 200

peak oil *see* oil, peak
Pearce, David 179
Pepper, David 33
permaculture 140, 170–2, 174
phenomenology 8, 69, 124, 222, 229–30, 232–4, 236, 241–3
  ecophenomenology 234, 237, 242
Phillips, Dana 7, 238
Pinkney, Tony 43
Pinkus, Karen 5, 181–2, 188
Platonism 130
Plumwood, Val 53–4, 58, 61, 63, 70n.1
Poe, Edgar Allan 160
Pohl, Frederik 161
Polak, Fred 159
pollution 43, 56, 77, 79, 88, 91,
populations
  animal 7, 57, 59–60, 97–104, 106–9, 113nn.1–4, 140–2
    see also extinction
  human xiii, 44, 56, 60–1, 146, 155, 163, 169, 195, 197–9, 200–1, 203
    see also Malthusianism
Posner, Richard 152
post-apocalypse 140, 150, 153, 159, 163, 202, 229
  see also apocalypse
posthumanism xiii–xiv, 4, 69, 208, 231
postmodernism 41, 182
Prestes, Fábio 86–9
progress 133, 177, 179, 181, 186–8, 200, 238
  economic 80
  environmental 166, 171
  scientific 198
Przybos, Julia 43–4

queering and queerness 111–12, 113n.6, 122, 181–2, 184, 186–92
  see also heteronormativity

Rawnsley, Andrew 34
Ray, Paul H. 196
realism 147, 163–4, 183, 233, 242
  see also magical realism
recycling 76, 79, 81–4, 93–4, 119, 219, 224
resilience 7, 54, 139–44, 147, 149, 156
rewilding 7, 57, 104, 108–10
Reynolds, Terry S. 13–14
Rigby, Kate 3, 6, 229
Rine, Abigail 184, 187–8, 191
Rio Earth Summit 77, 195
  see also United Nations
Robin, Libby 53
Robinson, Kim Stanley 161–2
  2312 163, 167
  Antarctica 163
  Mars trilogy 163, 167
  Orange County trilogy 163
  Science in the Capital trilogy 7, 163–74
  Shaman 163
  Years of Rice and Salt, The 163
Rockstöm, Johann 56
romanticisation 107–8, 236
Romanticism xiv, 33, 36, 42, 48, 62–3, 71n.7
Ronell, Avital 122
Rosa, Lúcia 83–4
Rose, Deborah Bird 53
Rousseau, Jean-Jacques 132
Rowland, Beryl 14
Rowley, William 18
Rubin, Larry 26
Ruskin, John 35

Santos, Eminéia Silva 84–5
satire 23–6, 147, 150, 197
Schroeder, Karl 161–2
Schumacher, E. F. 33
Schuster, Joshua 6–7, 59
science fiction (sf) 7, 128, 143, 145–7, 152, 159–64, 169, 174
Scigaj, Leonard M. 212–19, 222, 225
Searle, John 120
Sen, Amartya 13
sex 23, 111, 125, 146, 155, 185, 189, 202–6
sexuality 18–19, 105, 111, 131, 178, 181, 184
sf *see* science fiction
Shakespeare, William 19, 29, 117–18, 121, 125–6, 220
  Hamlet xiv, 118, 134n.1
  Henry IV Part 2 19
  Henry V 126
  King Lear 19

Macbeth 218
Tempest, The 219, 222
shallow ecology 53, 231–2
see also deep ecology; ecology
Shelley, Percy 33
simile 86
Singer, Peter 106
Smeaton, John 14
Snyder, Gary 104, 107, 219
socialism 33, 44
ecosocialism 161
Soper, Kate 41, 54, 61, 63
Sparrow, Tom 233, 242
speculative realism 7, 222, 229–30, 233–4, 242
Spinoza, Baruch 78
Spitzer, Leo 38–41, 43–4
Squire, Louise 7, 222
stability 107, 139–43, 147, 149, 155–6, 186
steady-state 39, 105, 112
Steffen, Will 55
Steinmüller, Karlheinz 160
Stiegler, Bernard 120, 123, 131
Stiglitz, Joseph E. 13
Stimson, Adrian 110
Stoekl, Alan 112
sublime and sublimity 7, 35, 115–16, 119, 123–7, 129–30, 132–3, 160
Sullivan, Heather I. 35
sustenance 38, 44, 46, 48–9, 93
Swift, Jonathan 196
synecdoche 85
Szeman, Imre 159

Taine, Hippolyte 44, 47
technology 21, 29n.7, 44, 48, 60, 103, 109, 111, 116, 121, 133, 146, 152, 155, 160, 169, 171–4, 175n.1, 177, 179, 185, 199
advances and innovations 14, 44, 59, 101, 159, 162
biotechnology 162
Tennyson (Lord), Alfred 19
theory, critical 6, 128, 133, 213, 229, 233
three pillars of sustainability 3, 6, 54–5, 76–8, 84, 94, 95n.1, 169, 179
see also triple bottom line
Thomas, Howard 2
Tiffin, Helen 239, 241
trans-corporeality 78, 88–9
transgender *see* gender
transnationalism 56, 68
see also nationalism
triple bottom line 52, 55, 77, 179
see also three pillars of sustainability
Tuana, Nancy 78–9, 87–8
Turley, Richard Marggraf 2
Turner, J. M. W. 14
Turner, Ted 107

Updike, John 207
Umwelt 37–8, 64–5, 69
United Nations (UN) 54–5, 77, 99, 178, 195, 198
see also Brundtland Report; Rio Earth Summit
utopia and utopianism 4, 33, 41–4, 47, 105–6, 140, 143, 147, 159–60, 162, 167, 169, 171, 173–4, 188–9, 192, 206
see also dystopia and dystopianism

van Dooren, Thom 107
Vaninskaya, Anna 42
Vendler, Helen 214–15, 221–2
Verne, Jules 160
Victorianism 33, 35, 42
Virgil 17
Vogt, Kristiina A. 211, 213, 219, 223, 225
von Uexküll, Jakob 37–8, 64

Wall, Derek 33, 35, 47
Walsingham, Thomas 24
Warminski, Andrzej 129–30
waste 76–7, 79–94, 106, 112, 119, 127, 154
wastefulness 218
water 13, 15, 17–18, 20, 22–3, 25, 27–8, 36, 42, 60, 71n.4, 91, 102, 159, 203, 221, 223, 239–40
management 26, 56
rights 22, 24–6, 198
security 13, 56
Watts, Martin 14
Wells, H. G. 152, 160

Westling, Louise 236, 238
Wheeler, Wendy 38
White, Hayden 144–5, 150–1
Williamson, Jack 160–1
Wilson, Eric 35
Winnett, Susan 181
Winterson, Jeanette 182, 187
  Boating for Beginners 182
  Oranges Are Not the Only Fruit 182
  Passion, The 182
  Sexing the Cherry 182
  Stone Gods, The 7, 177, 180–93
Wood, David 234, 236–7, 242–3
World Commission on Environment and Development, United Nations *see* Brundtland Report
Worster, Donald 33, 35

Zapf, Hubert 47
Zola, Emile 6, 41, 43–4, 47–8
  Fécondité 44, 46
  Germinal 43
  La Terre 34, 44–6, 49
  Thérèse Raquin 44
  Travail 44

EU authorised representative for GPSR:
Easy Access System Europe, Mustamäe tee 50,
10621 Tallinn, Estonia
gpsr.requests@easproject.com

www.ingramcontent.com/pod-product-compliance
Lightning Source LLC
Chambersburg PA
CBHW070631100726
47907CB00007B/1945

* 9 7 8 1 5 2 6 1 8 2 3 5 7 *